ENGINEERING SEISMOLOGY AND
EARTHQUAKE ENGINEERING

Library of Congress Catalog Card Number: 73-85150

ISBN-13. 978-94-011-7576-0   e-ISBN-13: 978-94-011-7574-6
DOI: 10.1007/ 978-94-011-7574-6

© 1974 Noordhoff International Publishing, a division of A. W. Sijthoff International
Publishing Company B.V.

Softcover reprint of the hardcover 1st edition 1974

ENGINEERING SEISMOLOGY AND EARTHQUAKE ENGINEERING

edited by

JULIUS SOLNES

NOORDHOFF — LEIDEN — 1974

NATO ADVANCED STUDY INSTITUTES SERIES

Proceedings of the Advanced Study Institute Programme, which aims at the dissemination of advanced knowledge and the formation of contacts among scientists from different countries.

The series is published by an international board of publishers in conjunction with NATO Scientific Affairs Division

A	Life Sciences	Plenum Publishing Corporation
B	Physics	London and New York
C	Mathematical and Physical Sciences	D. Reidel Publishing Company Dordrecht and Boston
D	Behavioral and Social Sciences	Sijthoff International Publishing Company Leiden
E	Applied Sciences	Noordhoff International Publishing Leiden

Series E: Applied Sciences - No. 3

PREFACE

by Júlíus Sólnes

An Advanced Study Institute on engineering seismology
and earthquake engineering was held in Izmir, Turkey
July 2-13, 1973 under the auspices of the Scientific
Affairs Division of NATO.

The Institute was organized by an organizing
committee headed by the two scientific directors and
with representation by the Turkish National Science
Foundation, Turkish National Committee for Earthquake
Engineering, the Middle East Technical University and
the Aegean University.

93 scientists and engineers of 18 countries took
part in the work of the Institute which comprised 10
working days with lectures, discussions and panel
meetings.

The main lecture topics of the Institute were
covered in five main sections:
1. Generic causes of earthquakes.
2. Ground motion and foundation response.
3. Earthquake response of structures and design considerations.
4. Codes and regulations; implementation.
5. Earthquake hazards and emergency planning.

Upon completion of each section, general discussion
and short presentations by several of the participants
took place and summary statements were offered by the
main lecturers. The atmosphere of the meetings was in-

formal and cordial thus giving rise to many unorthodox
and newly conceived ideas.

Several of the main lectures have been selected for
publication as presented herein. It should be understood that more emphasis has been put on material for
advanced instruction rather than presentations of
recent research projects, in corcordance with the purpose and scope of the Institute. Therefore, it is believed that this book offers suitable textbook material
for an advanced study into the various aspects of
earthquake engineering and engineering seismology.

Besides thanking all the lecturers for their contribution to this book and all the participants for their
enthusiasm which made the above institute successful,
I would like to express my sincere thanks to Mrs. Kaja
Svendsen for her patient and careful typewriting of all
the papers and to Benny Leisten and Christian Bramsen
who between themselves solved the often impossible job
of producing usable figures for the papers. All this
work was carried out while we were working together at
the Structural Research Laboratory of the Technical
University of Denmark.

Reykjavík, March 1974.

TABLE OF CONTENTS

VIII

SEISMOTECTONICS OF THE EASTERN MEDITERRANEAN AREA

by   B.C.Papazachos

National Observatory of Athens
Seismological Institute
Athens - Greece

ABSTRACT

    The main geophysical properties of tectonic signi-
ficance of the eastern Mediterranean and surrounding
area are briefly reviewed. An effort has been made to
further advance the idea that the seismic activity in
this area is mainly the result of relative movements of
aseismic lithospheric blocks. The best of the available data as to the distribution of earthquake foci and
focal mechanisms have been used to determine the zones
of thrusting or extension, as well as the transform
faults which define the boundaries of the most important
aseismic blocks in this area. The directions of the
relative motions between these blocks have been estimated. Most of the seismic and other geophysical properties of the area are in accord with this seismotectonic model. However, there are several observations
which show that the tectonics of this area are too complicated to be fully understood by simple seismotectonic models. Research based on more accurate new seismic data as well as on other geophysical and geological
information must be carried out for a better understanding of the seismotectonic process in this area.

INTRODUCTION

The Mediterranean sea fills the long depression which extends from the Strait of Gibraltar eastward up to the Lenanese coast. It is believed that the area which is now occupied by the Mediterranean was the westernmost part of a once-vaster oceanic area, called Tethys. Tethys stretched eastward through what is now the Middle East and Asia. This oceanic area closed during the opening of the Atlantic ocean. During the last stage of this process Africa and Eurasia approached each other creating compressive forces in an approximate north-south direction. These forces produced the Alpine mountain range. It was formerly believed that the Medeterranean sea was a direct remnant of Tethys, but there is evidence that the Mediterranean is relatively young and therefore cannot be considered as such (Smith[1]).

The eastern and western Mediterranean are separated by the Calabrian arc. Striking differences as to several geophysical properties exist between the two sections (Papazachos.[2]). It is suggested that the present basin of the eastern Mediterranean is a fragment of a younger ocean created by tearing off microcontinents from north Africa (Rabinowitz and Ryan.[3], Dewey et al.[4]). On the other hand, evidence has been presented that oceanization may take place in the western Mediterranean (Ritsema.[5]).

The western part of the eastern Mediterranean sea is called the Ionian sea while its easterly part is occupied by the Levantine basin. Branches of the eastern Mediterranian are the Adriatic sea between Italy and Yugoslavia-Albania and the Aegean sea between Turkey, the mainland of Greece and the island of Crete. Recent tectonic activity is considerable in the eastern Mediterranean and increases from south to north. Several geophysical and geological observations are evidence of this activity. The seismicity, gravity anomalies, volcanic activity, deformation of sea sediments as well as the folding and thrusting of the young mountain ranges which bound this area northward, are among the most important of these observations.

The eastern Mediterranean area is part of the continental fracture system of the earth and for this reason it belongs to areas where thrusting and crustal destruction take place. However, recent magnetic and seismic work has shown that small zones of extension, where new crust is probably being formed, exist in this area.

The investigation of the seismotectonics of an area is based on two main kinds of seismic data: data concerning the spatial distribution of the earthquake foci and on data concerning the focal mechanism of the earthquakes. Information on the distribution of earthquake foci has existed for a relatively long period, but data on the focal mechanism are available only for the last twenty-five years or so.

Information on seismic destructions in this area has existed for a long time, as this area has been inhabited by civilized people since early times, but it was only in the last half of the nineteenth century or at the beginning of the present century that the first attemps were made to conduct a systematic investigation of the seismic activity in certain regions of the land surrounding the eastern Mediterranean (Mallet[6], Schmidt[7], Ballore.[8]).

Interesting work on the seismicity in several regions of the eastern Mediterranean area and its relation to the tectonics was done in the first half of the present century (Sieberg,[9] Mihailović,[10] Magnani.[11]). However, most of this work has been based on non-homogeneous macroseismic data and on rather confused ideas regarding this relation. The systematic publication of information as to the location, time and magnitude of earthquakes by several seismological centers (ISS, BCIS and USGGS), and the collection of additional and more accurate macroseismic data contribute much to the investigation of the seismicity of several regions of this area and its tectonic implication (Pinar,[12] Ketin and Roesli,[13] Galanopoulus,[14],[15,16] Ergin.[17]).

The publication of catalogues containing new data and more homogeneous revised old data (Gutenberg and Richter,[18] Ergin et al,[19] Karnik,[20] Papazachos and Comninakis.[21]), contribute to a further understanding of the seismotectonics of this area (Karnik,[22] Caputo et al,[23] Papazachos and Comninakis,[24] Papazachos.[25]).

Determination of the focal mechanism or of the fault plane solution, as it is usually called, of earthquakes in the eastern Mediterranean area has been made for many earthquakes by several seismologists (Hodgson and Cock,[25] Papazachos,[26] Ocal,[27, 28] Canitez and Ucer,[29] Wickens and Hodgson,[30] Delibasis,[31] McKenzie,[32] Canitez.[33]). Attempts have been made for statistical treatment of such solutions with the purpose of obtaining information regarding the tectonics of this area (Scheidegger,[34] Constantinescu et al,[35] Canitez and Ucer,[36] Shirokova,[37] Papazachos and Delibasis.[38]).

The fast promotion of the ideas of the new global tectonics and synthetic work based on data on foci distribution and focal mechanism, as well as an other geophysical and geological information, has much advanced our knowledge of the tectonics of this area (Ritsema,[39,5] McKenzie,[40,32] Lort,[41] Comninakis and Papazaghos,[42] Finetti and Morelli,[43] Dewey et al.[4]).

The main purpose of the present work has been the use of the more reliable published data on the distribution of the earthquake foci and focal mechanism in the eastern Mediterranean and the surrounding area to study the seismotectonics of this area in more detail than heretofore. Geophysical information is also used to support the conclusions based on the above-mentioned seismic data in some regions, or to emphasize the necessity for additional and more accurate data in other regions. A general description of the psysiographic and of several geophysical properties of tectonic significance is first given.

PHYSIOGRAPHIC FEATURES OF TECTONIC ORIGIN

The dominant physiographic features of tectonic origin in the eastern Mediterranean are the Hellenic and Calabrian arcs, the Mediterranean ridge and several trenches and throughs. Several abyssal plains, cones etc. have also been observed (Fig.1).

The Hellenic arc, separating the Aegean from the Mediterranean is defined by the Hellenides, the Ionian islands. Peleponnesus and the Kythera, Crete, Karpathos and Rhodes islands. A continuation of this arc to the west is an arcuate structure formed by the western Taurus. Mountains in south Turkey and by Cyprus. The Hellenic arc has the main properties of a typical island arc (Papazachos and Comninakis.[24]). In its inner part and at a mean distance of about 120 km , the sedimentary arc is paralled by a volcanic arc, which consists of volcanic islands. It contains the volcanoes of Methana, Santorine and Nisyrus as well as sever-

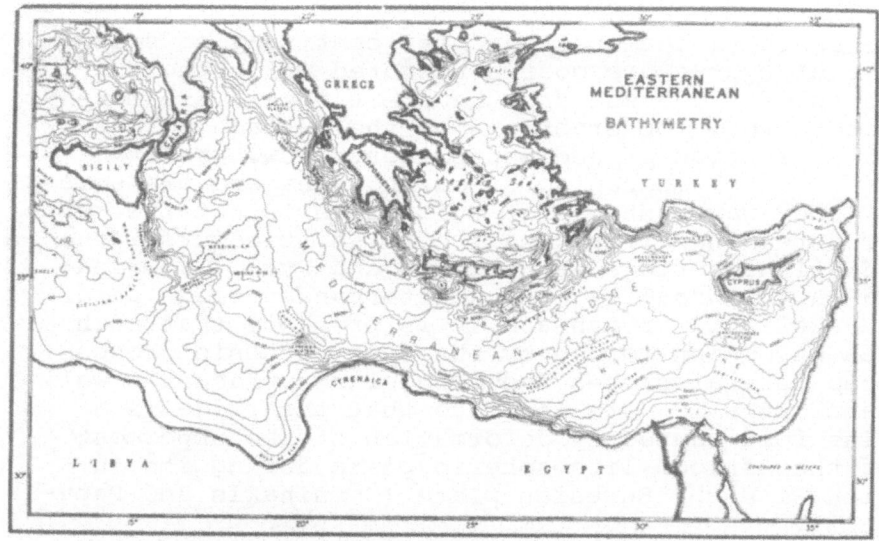

Figure 1:   Physiographic Features of the Eastern
             Mediterranean Area.

ral solfatara and fumarole fields (Georgalas.[44]). Between Crete and the volcanic arc a trough extending in
an east-west direction exists. In this trough the
maximum water depth, which is also the maximum depth in
the whole Aegean province, is about 2000 m. The
second maximum depth in the Aegean sea has been measured
in a trough which exists in the northern Aegean and extends in a northeast-southwest direction.

    In the convex side of the Hellenic arc there are
relatively deep oceanic trenches, which parallel the
arc and form a system usually called the Hellenic
Trench. The water in some of these trenches has a
depth of between 4000 m and 5000 m. The Ionian trench
near the western coast and the Pliny and Stravo
trenches southeast of Crete are the most well known of
these trenches. The structure between the Hellenic arc
and the Mediterranean ridge, which includes the linear
trenches and the depressions that parallel the arc, is
called the Hellenic Trough and has a length of more
than 1500 km (Ryan et al.[45]).

    The Mediterranean ridge, recently discovered in
the mid-eastern Mediterranean sea, extends from the
northern Ionian sea to Cyprus, has a length of about
1600 km and averages about 150 km wide (Emery et al[46]).
The median line of this ridge is not marked by a rift
valley but separates areas of different textures. The

surface south of the crest and its continuation south
and east of Cyprus are coarse-textured while the surface north of the crest and the crest itself are more
finely textured. Observations on the surface morpho-
logy, distribution of sediments, heat flow, gravity,
magnetics, focal mechanism of earthquakes and transmission of body waves show that this ridge is not the result of spreading of the sea floor, as it occurs with
the mid-Atlantic ridge (Rabinowitz and Ryan, [3] Comninakis and Papazachos.[42]). The Mediterranean ridge parallels the Hellenic trench and it is probable that both
features have been formed by the same tectonic process.
Its steep lateral slopes and irregular surfaces as well
as several seismic data indicate that the ridge is a
submarine fold caused by deformation of the uppermost
part of the African lithospheric plate during its interaction with the Eurasian plate (Comninakis and Papazachos.[42]).

There are not extensive abyssal plains in the eastern Mediterranean as there are in the western Mediterranean. The largest of these plains is the Herodotus
abyssal plain on the south side of the Mediterranean
ridge. There is evidence that this plain is a remnant
of a once-vaster area of sedimentary accumulation (Ryan
et al.[45]). South of the Herodotus abyssal plain is the
Nile cone which spreads out from the Nile delta.
The Calabrian arc is formed by Calabria, Sicily,
the Strait of Sicily and the Atlas mountains and it separates the eastern from the western Mediterranean. No
deep topographic trench on the convex side of the arc
is observed but seismic reflection profiles, across the
Messina cone, reveal a buried depression now filled by
sediments (Rabinowitz and Ryan. [3]).

GEOPHYSICAL PROPERTIES OF TECTONIC SIGNIFICANCE

A brief description of the gravity, magnetic and
thermal properties of the eastern Mediterranean follows.

The gravity field in the eastern Mediterranian is
characterized by low intensity values over most parts
of the region on the convex side of the Hellenic arc.
The most prominent features of the gravity field are
two belts of negative free-air anomalies along the Hellenic trough and the Mediterranean ridge. The values
of the free-air anomalies in the southern belt, which
is associated with the Mediterranean ridge, range between -250 mgal , in the west, and -150 mgal, in the

east, while values in the northern belt, which is associated with the axis of the Hellenic trough, range between -120 mgal, in the west, and values more than
-230 mgal over the deep trough southeast of Rhodes
(Rabinowitz and Ryan.[3]). These belts of negative anomalies are probably related to deep crustal structure
and indicate the isostatic inbalance present in the
area (Woodside and Bowin.[47]). Another belt of negative
free-air anomaly has been observed south of Calabria.
This is probably related to the Calabrian arc. Positive free-air anomalies are found on Cyprus, in the
Aegean sea, over the Nile and in the western Ionian sea.
The maximum positive anomalies have been observed in
Cyprus (+280 mgal) and in the Agean (+120 mgal). The
gravity anomalies in Cyprus are associated with the ultrabasic Troodos massif.

The Bouguer anomalies in the eastern Mediterranean
arc predominantly positive, as might be expected for an
oceanic area, but are substantially less positive than
Bouguer anomalies in the western Medeterranean or in
deep oceanic areas in general (Woodside and Bowin.[47]).
Values greater than +150 mgal have been obtained only
on small isolated portions of the eastern Mediterranean.

Harrison[48] deduced from gravity data that a zone
of compression exists along the Hellenic arc and a
fault dipping to the north occurs at the junction between the shallow Aegean and the deaper sea to the
south of Crete. This conclusion was later confirmed by
data on focal mechanism (Papazachos and Delibasis,[38])
and by data on the distribution of earthquake foci (Papazachos and Comninakis.[49]).

Rabinowitz and Ryan[3] found a dependence of the
free-air anomalies on the topography across the Hellenic trench similar to the dependence of this anomaly on
the topography in other well-known trenches of the
earth. They observed similar variation of the gravity
field intensity over the buried depression south of the
Calabrian arc.

Measurements of the total magnetic field intensity
in the eastern Mediterranean (Vogt and Higgs,[50] Allan
and Morelli.[51]) reveal that the Hellenic trough, the
Mediterranean ridge, the Hellenic arc, the trough north
of Crete and the Anaximander mountains between Cyprus
and Rhodes are magnetically undisturbed. Magnetic anomalies have been observed in Cyprus, the Aegean sea and
the southwestern Ionian. A distinct arcuate magnetic

high has been observed over the igneous ultrabasic in-strusive on Cyprus. This anomaly continues into the Turkish mainland to the northwest. In the Aegean the magnetic anomalies are characterized by dominant north-east-southwest trends. These anomalies are more in-tense in the northern Agean and parallel the structural units of the Hercynian granites.

Magnetic lineations, similar to those found over major oceans, have not been detected in the eastern Me-diterranean and for this reason the recent movements of the crust cannot be dated by magnetic observations. This is due to the fact that the Mediterranean is the site of shrinking. However, the intense magnetic anom-alies, observed in the northern Aegean sea, suggest the existence of a tensional graben, along which magnetized rocks have been intruded (Vogt and Higgs.[50]). This in-terpretation is supported by seismic observations (McKenzie.[40]).

Fourty heat flow measurements have been made at places in the Mediterranean ridge, in the Hellenic trench, in the abyssal plains and in the vicinity of the region of large magnetic anomalies around Cyprus. The measured values are nearly the same in all four provinces. The average of these values is equal to 1.00 HFU (1 cal/cm$^2 \cdot$sec) with small standard devia-tions. This is significantly lower than the world-wide average of 1.35 HMU for oceanic provinces and about half of that in the western Mediterranean (Ryan et al.[45]).

SEISMOTECTONIC MODELS

Several attempts have been made to interprete the distribution of the earthquake foci and the focal mech-anism of the earthquakes in the Mediterranean area. Two main models to interpret the seismic activity in this area have been proposed by McKenzie and Ritsema.

McKenzie[40],[32] suggests that the relative motion between several lithospheric plates is the main cause of the seismic activity in the Mediterranean ares. Fi-gure 2 shows these plates and the directions of their motions. The plates with numbers (1), (2) and (8) are the big Eurasian, African and Arabian lithospheric plates, respectively. Those with numbers (5), (6) and (7) are relatively small plates and he calls them the Turkish, Aegean and Black Sea plates, respectively. No name has been given to the plate which is formed by

northern and central Greece and by the northernmost
part of the Aegean Sea. The plates are separated by
thrust zones and zones of extension, which are denoted
by thick lines crossed by small normal lines and by
double lines respectively, as well as by transform
faults. According to McKenzie, the fast motion of the
Aegean and Turkish plates relative to Africa and Europe
can account for most of the seismic activity in this
area.

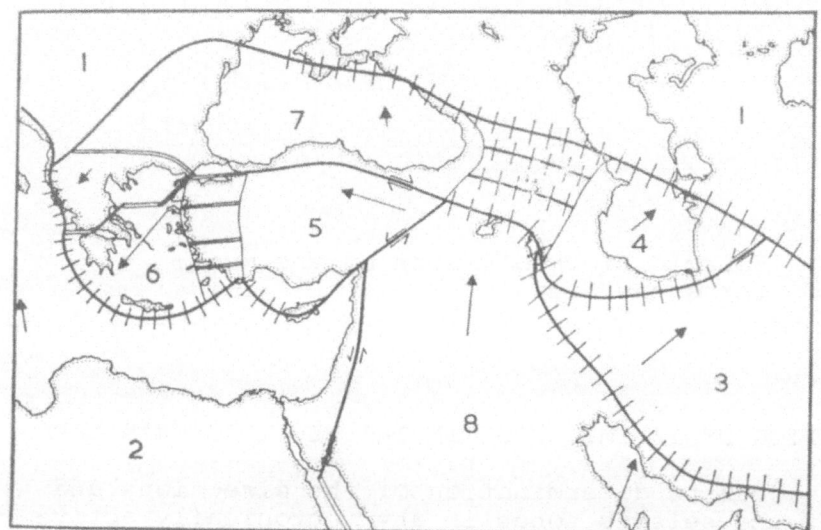

Figure 2:   Motion of Lithospheric Plates in the
            Eastern Mediterranean Area.

        Ritsema [39],[5] used seismic data and some geophysical
and geological information to investigate the tectonics
of the Mediterranean area. He has concluded that drift
of plates is not the only active agent in this region.
Passive gravity sliding and flow of mantle material in
the low velocity layer of the mantle are likely to play
an important role. He has presented evidence that a
stress field and relative movements of arcs, blocks or
plates in an approximate west-east, east-west direction
can explain the distribution of basins and ranges in
the Mediterranean region. Figure 3 shows the seismic
division of the Mediterranean area in blocks (I to VI),
their motions (arrows) relative to blocks (II), the
active boundary lines (1 to 10) of relative displacement, basins deeper than 2000 m (a to g) and the
Agean shallow basin (h).

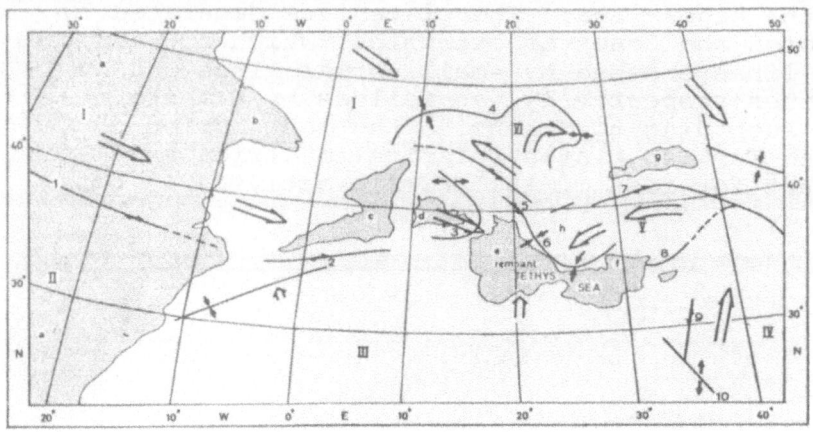

Figure 3:   Seismic Block Division of the Medi-
            terranean Area.

DISTRIBUTION OF SEISMIC FOCI IN THE MEDITERRANEAN AREA

The accurate determination of the directions and
slopes of the seismic zones in any tectonically active
area is of great importance since these zones define
the boundaries of aseismic blocks or lithospheric
plates. Several big lithospheric plates of the earth
have been determined by this method.

The problem is difficult when we want to define
the boundaries of small aseismic blocks like those
existing in the eastern Mediterranean area because not
all the foci of the known earthquakes have the accuracy
required for such determination.

The published focus coordinates of the earthquakes
which occurred in the last decade are, in general, more
accurate than the old ones. Nevertheless, there are
zones where few or no earthquakes have recently occurred
but it is known from older reliable observations that
earthquakes, and in places large ones, have occurred in
these zones. On the other hand, even the foci of the
earthquakes of the last decade are not known with sufficient accuracy when their magnitudes are small. For
this reason, the use of the new observations only in

determining boundaries of small aseismic blocks in the
area studied in the present paper can be misleading. It
is therefore necessary to increase the data sample by
including reliable information concerning older earthquakes. To do this we must take advantage of the fact
that the foci of the larger earthquakes of a certain
time period are more accurately known than the foci of
smaller earthquakes of the same period just because the
distribution of the recording seismic stations and the
macroseismic effects depend greatly on the magnitude of
the earthquakes.

The present author Papazachos[2] has determined the
main seismic zones of the shallow shocks in the Mediterranean and the surrounding area by plotting epicenters in such a way as to make allowance for the magnitude and year of occurrence. The time between 1901 and
1971 has been properly divided into four intervals and
a minimum magnitude has been chosen for each interval.
From the earthquakes of each interval only those with
magnitudes equal to, or larger than, that chosen have
been considered. The selection of the time intervals
and the smallest earthquake magnitude of each interval
has been made by taking into account the progress in
the establishment of regional and global networks of
seismic stations and in the methods used to determine
earthquake foci.

Figure 4 shows the distribution of the epicenters
of shallow earthquakes occurring in the Mediterranean
area between 1901 and 1972. The seismic zones are denoted by dashed lines. The more active and certain of
these zones are shown by thick lines while the less
active or less certain are denoted by thin lines. Although there is much scattering in the distribution of
the epicenters, several aseismic blocks are distinguishable. The western Mediterranean forms one aseismic
block and the Adriatic sea forms another block. Two
blocks are distinguished in the eastern Mediterranean
sea: the Ionian block in the west and the Valentine
block in the east. The Aegean blok is part of what
McKenzie[32] calls the Aegean plate. The part of Turkey,
shown in Figure 4, is formed by three blocks. The
northernmost of these is the northern Anatolian block.
The block in western Turkey is small and it may not
exist, but the block in southern Turkey is well defined.
Other aseismic blocks are also distinguishable.

Figure 5 shows epicenters of non-shallow ($h \geq 60$ km
earthquakes which occurred in the Mediterranean area

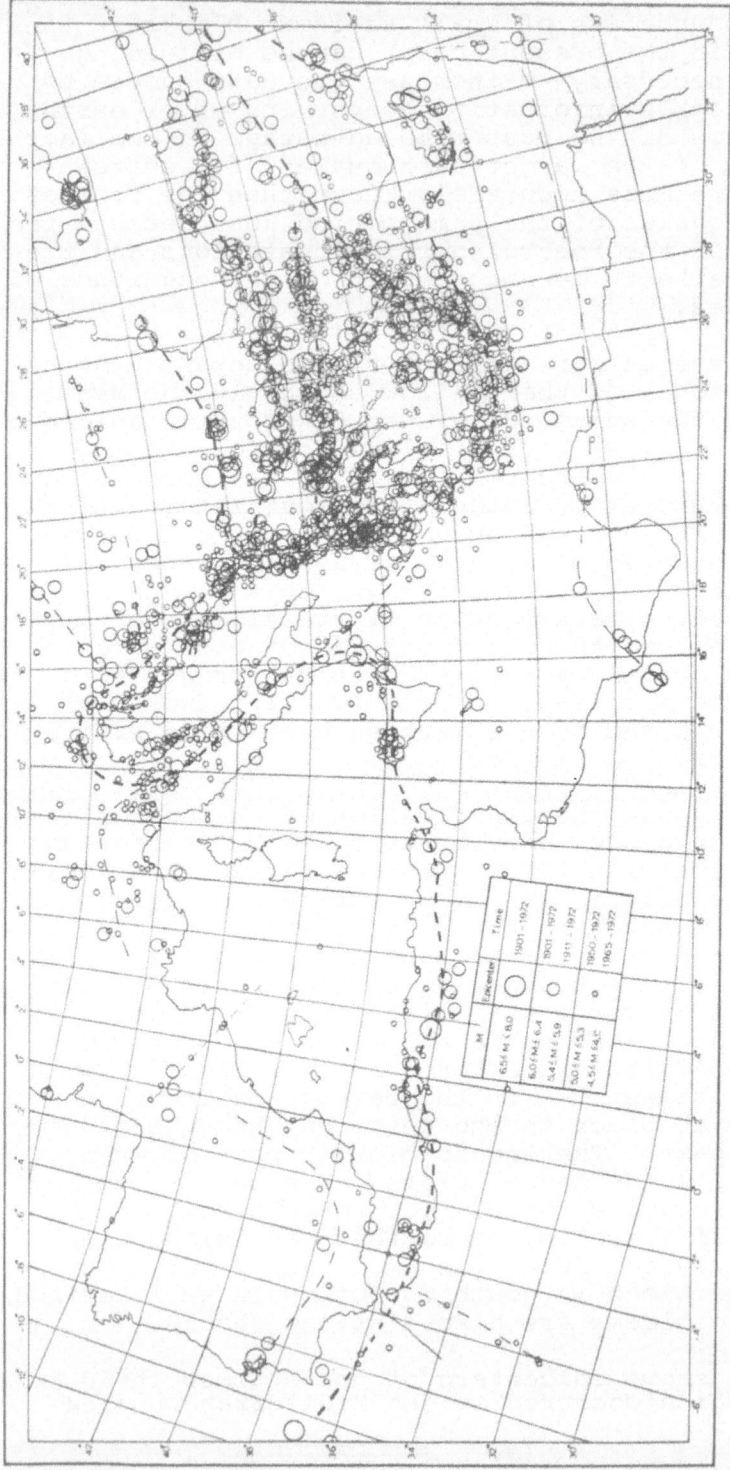

Figure 4: Distribution of Epicenters of Shallow Earthquakes Occurring in the Mediterranean Area Between 1901 and 1972.

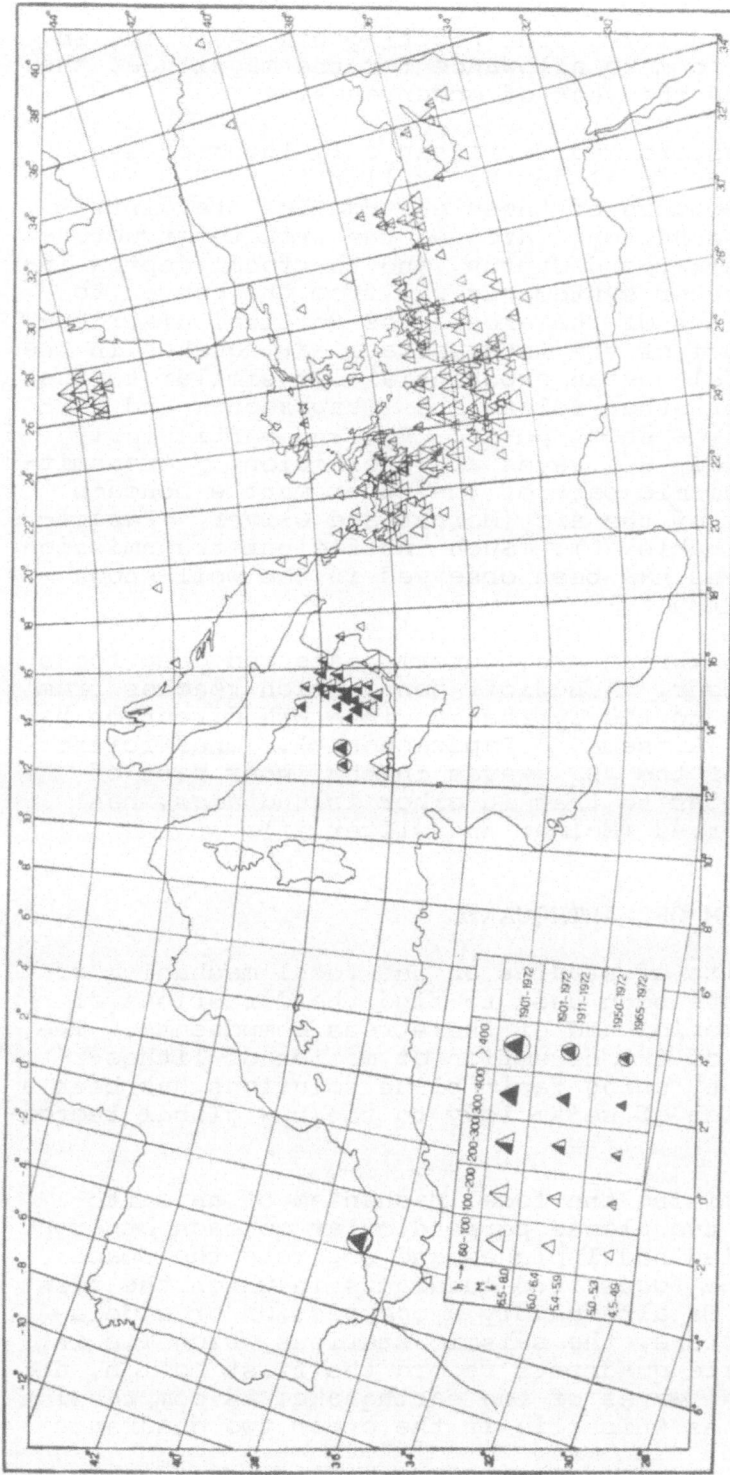

Figure 5: Distribution of Epicenters of Non-Shallow Earthquakes (h>60 km) Occurring in the Mediterranean Area Between 1901 and 1972.

between 1901 and 1972. The plotting has been made in
such a way as to make allowance for the magnitude, the
focal depth and the year of occurrence.

In the eastern Mediterranean only intermediate
earthquakes with focal depths smaller than 200 km occur. The epicenters of these earthquakes are distributed along the Hellenic arc and the arcuate structure
in southern Turkey and Cyprus, and the focal depths increase in a rather systematic way from the convex to
the concave sides of the arcs. The vertical distribution of the foci of the intermediate earthquakes in the
Hellenic arc follows an exponential law similar to that
which holds for other island arcs (Papazachos and Comninakis.[24]). The short period P , and particularly
the short period S , waves are inefficiently transmitted in the aseismic part of the upper mantle beneath
the inner part of the arc (Molnar and Oliver[52] Papazachos and Comninakis.[24]). Such inefficient transmission
of seismic waves has been observed in the well-known
island arcs (Utsu.[53]).

In the Calabrian arc, intermediate and deep focus
earthquakes occur. A Benioff zone, which reaches from
the Ionian sea to the Tyrrhenian in a WWN direction has
been observed (Ritsema,[54] Papazachos.[2]). Inefficient
transmission of the S_n waves in the inner part of
this arc, similar to that in other island arcs, has
been also observed (Molnar and Oliver.[52]).

FOCAL MECHANISM OF EARTHQUAKES

The results of studies on the focal mechanism of
earthquakes have been used to find the directions of
the tectonic motion and of the stress components. The
determination of the direction of motion of lithospheric plates by using fault plane solutions has been a
big contribution of seismology to the new global tectonics.

In determining the focal mechanism of an earthquake we find two planes perpendicular to each other,
which are called nodal planes, and separate the space
surrounding the focus into quadrants in which the first
motion of P is alternately a compression or a delatation. Therefore, the seismic stations which lie in
the two opposite quadrants record the first motion, due
to longitudinal waves of the earthquake, as compression
and the stations which lie in the other two quadrants

record this motion as delatation. The intersection of
the two planes is called axis B . The one of the nodal planes is the fault plane. The motion takes place
along an axis which lies on the fault plane and is
usually called kinematic axis C . The other nodal
plane is normal to the direction of motion and it is
called the auxiliary plane. The axis which lies on the
auxiliary plane and is normal to the fault plane at the
focus of the earthquake is called axis A . The angles
of the axes A and C are bisected by two other axes,
which are denoted by P and T . The axis P lies in
the quadrants which contain delatations and the axis   T
in the quadrants which contain compressions. It is
easy to see that the axis B is normal to the plane
which contains the other four axes. Where the slipping
does not occur along a pre-existing fault and the material in the focal region is homogeneous, the axes P,
T and B have the directions of the maximum compression,
the maximum tension and the intermediate stress component, respectively. For this reason the results of the
focal mechanism studies have been used to determine the
directions of the three stress components at the focal
regions of earthquakes.

The focal mechanism studies or the determination
of fault plane solutions are usually based on the first
onset of the P waves as it is recorded by several
stations but, in some cases, data concerning S or
surface waves have also been used.

McKenzie[55] has theoretically shown that the axis
of maximum compression can be anywhere in the quadrant
containing the P axis, when the slip takes place along
pre-existing weak planes. Since the shallow earthquakes occur along weak zones which exist in the boundaries of the lithospheric plates, it is not possible to
determine the directions of the stress components by
fault plane solutions. However, several authors still
use fault plane solutions of shallow earthquakes to determine the directions of the stress components. Such
determinations are in agreement with the expected
stress direction in some regions (Maasha and Molnar,[56]
Comninakis and Papazachos [42]) or with the direction of
stress determined by other methods (Sykes et al.[57]).
Some success of the method is also admitted by McKenzie
who attributes it to the fact that the P axis varied
by 45° from the slip direction which is invariant in
a certain region.

We have examined all the published fault plane so-

lutions of the earthquakes which occurred in the eastern Mediterranean and surrounding area between 1948 and
1969 with the purpose of distinguishing the most reliable of them. Such solutions have been found for 70
earthquakes. Only earthquakes which have magnitudes
larger than 6 and occurring between 1948 and 1963 or
have magnitudes larger than 5.5 and occurring after
1963 when the network of standardized stations were in
regular operation have been considered. The information
concerning the focal mechanism of 13 earthquakes has
been taken from Canitez and Ucer,[29] of 25 from Papazachos and Delibasis,[38] of 31 from McKenzie[32] and of
one from Canitez.[33].

DIRECTIONS OF P AND T AXES

Although the relation of the P and T axes to the
stress components is not clear in the case of shallow
shocks, a proper plot of these axes can at least give
useful information regarding the kind of faulting.

The directions of the P and T axes for the earthquakes mentioned above are plotted in Figure 6. The
number close to each epicenter is the last two figures
of the year of occurrence. In the cases when more than
one of these earthquakes have the same epicenter only
the P or T axis of the earthquake with the best fault
plane solution is shown in Figure 6. For some places
some additional data, concerning smaller earthquakes
have been used, because for these places no other data
exist. In these cases the year of occurrence of each
earthquake is written close to the symbol. In the cases
when one axis makes a small angle (<25°) with the
horizontal plane, and the other axis makes a large
angle (>45°) with the same plane, the axis making a
small angle with the horizontal plane is denoted by a
long thick symbol while the other axis is denoted by a
short thin symbol. In the other cases the two axes are
denoted by thin symbols which have lengths proportional
to the cosines of the angles between the axes and the
horizontal plane.

It is understood that the long thick T symbol
means mainly normal or extensional faulting and the
long thick P symbol means mainly thrust or reverse
faulting. In the cases when both symbols are thin, the
faulting is mainly strike slip. According to the ideas
of the new global tectonics, separation of the blocks
or plates takes place in the first case while in the

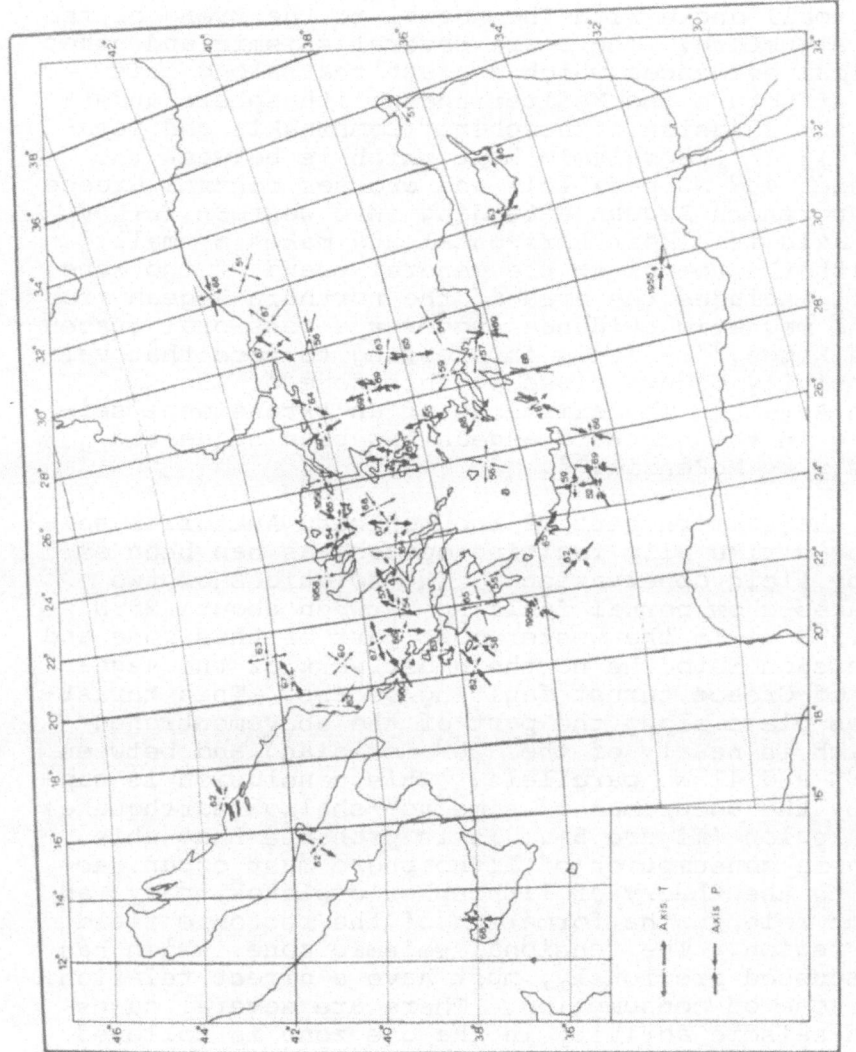

Figure 6: Direction of P and T axis for Eastern Mediterranean Earthquakes Occurring Between 1948 and
1969.

second case the two blocks move towards each other and consumption of lithosphere takes place. In the last case slip between plates occurs along transform faults.

We can observe in Figure 6 that along the convex side of the Hellenic arc south of the 38° N parallel and in Cyprus the axis P is horizontal and in general makes a small angle with the normal to the trend of the arcuate structure. There are several seismic and other geophysical evidences which suggest that along this arcuate structure the Mediterranean lithosphere under-thrusts the Eurasian lithosphere (Comninakis and Papa-zachos.[42]). In the seismic belt which is between the 38° N and 40° N parallels and crosses central Greece and the northern Aegean extending into western Turkey the T axis is almost horizontal and makes a small angle with the normal to the general trend of the zone. This belt includes the area of the northern Aegean, for which the magnetic evidence suggests a tensional graben Vogt and Higgs.[50] It is interesting to note that vol-canic activity occurs close to this zone of the northern Aegean. The existence of an extensional sei-smic zone in the northern Aegean has been suggested previously by McKenzie.[40].

In the eastern part of the northern Anatolian sei-smic zone strike slip faulting occurs, as has been ex-pected by field observations. The solutions of two earthquakes show nornal faulting between about 28°N and 30.5°N . In the westernmost part of this zone and its extension into the northernmost part of the Aegean Sea and of Greece thrust faulting occurs. This thrust-ing takes place along the part of the abovementioned zone which is nearly of the 28° meridian and between the 40°N and 41°N parallels. This conclusion is sup-ported by the occurence of some non-shallow earthquakes in this region (Figure 5). It is probable that this zone, where consumption of lithosphere must occur, ac-cording to the theory of lithospheric plates, plays an important role in the formation of the tectonic field in this region. The tensional seismic zone, which has been discussed previously, must have a direct relation to this zone of consumption. There are several cases in which seismic activity in the one zone is followed by seismic activity in the other zone within a rela-tively short time. As an example we can mention the earthquake sequence which occurred in northwestern Tur-key in the spring of 1969. McKenzie[32] has determined the fault plane solution for six earthquakes of this sequence by using the long-term records of standardized stations. The first of these earthquakes occurred in the northern thrust zone on March 3, 1969 and was caused by reverse faulting. The second earthquake occurred in the tensional zone by normal faulting on March 23, 1969. The other four earthquakes have also occurred in the tensional zone by normal faulting.

The southwesternmost part of Turkey is characterized by normal faulting. The tectonic explanation of these observations is difficult and is discussed in the next sections.

Some fault plane solutions of intermediate earthquakes in this area have been determined by several of the authors mentioned above. What is of tectonic importance in the fault plane solutions of the intermediate and deep focus earthquakes is the direction of P or T axis (Isacks et al.[58]).

The occurrence of a non-shallow earthquake can be considered as a failure of a homogeneous material with little internal friction (McKenzie.[55]). For this reason the P and T axes show the directions of the stress components at the foci of these shocks. The T axes of the intermediate earthquakes in the Hellenic arc dip approximately along the descending lithospheric slab, as it occurs in other island arcs where no deep focus earthquakes occur (McKenzie.[32]).

SLIP DIRECTIONS

In trying to determine the direction of fault slipping by using fault plane solutions, we meet with great difficulty because there is no reliable and generally applicable method to distinguish the fault plane from the auxiliary plane or its equivalent, to find which is the axis C along which the slip takes place, and which is the normal to the fault plane axis A. Some criteria used by several authors to resolve the ambiguity are rather weak. In some areas where the faulting is shown on the surface there is no problem. In some other cases the distribution of the aftershocks can be used to resolve the ambiguity. A way to find a solution to the problem in some areas is to make use of corollaries of a known theory or hypothesis as is the theory of plate tectonics. According to this theory, the motion along an arc must have such a direction as to make the two plates converge and the oceanic plate underthrust the continental one.

If we make the assumption that the theory of the lithospheric plates, which has been derived for large plates, is directly applicable to this region, where relatively small aseismic blocks exist, the distinction between fault planes and auxiliary planes in the thrust zones can be made. This distinction is also possible for the earthquakes in the northern Anatolian seismic zone, since field observations have shown that dextral faulting occurs there (Ketin,[59] Ambraseys and Zatopek.[60]). In some cases the distribution of aftershocks or macroseismic data have been used to resolve the ambiguity. There are some earthquakes, particularly those in the inner part of the Hellenic arc, for which the decision is very difficult.

Figure 7 shows the tectonic feature of the seismic zones in the eastern Mediterranean area as they have been determined by the horizontal and vertical distribution of the earthquake foci as well as by fault plane solutions. The zones of thrusting and tension are shown by thick lines and two parallel thin lines, respectively, while the transform faults are denoted by single thin lines. The dashed lines denote zones for which the tectonic features cannot be reliably determined by the available seismic data. In the same figure the horizontal projections of the slip vectors are shown at several places along the seismic zones. The direction of each vector is the mean horizontal direction of the slip vectors determined by fault plane solutions of the shallow earthquakes which have epicenters close to the place where each vector is shown in this figure. The length of each vector is proportional to the cosine of the mean angle between the slip vectors and the horizontal plane.

While the data show that the eastern Mediterranean lithosphere underthrusts the Eurasian lithosphere along the arcuate structure between the Ionian islands and Cyprus, it is not clear in what direction the underthrusting takes place along the thrust zone which crosses the northernmost part of the Aegean sea.

The extensional zone which crosses the northern Aegean is related to both zones of thrusting. Three tensional zones surround a block in western Turkey. The slip vectors converge to the center of the block as if a zone of subduction existed there. There is some evidence, although weak, to support it. For an earthquake which occurred there ($38.2^\circ N$, $29.7^\circ E$) on July 9, 1933, a focal depth equal to 100 km has been determined

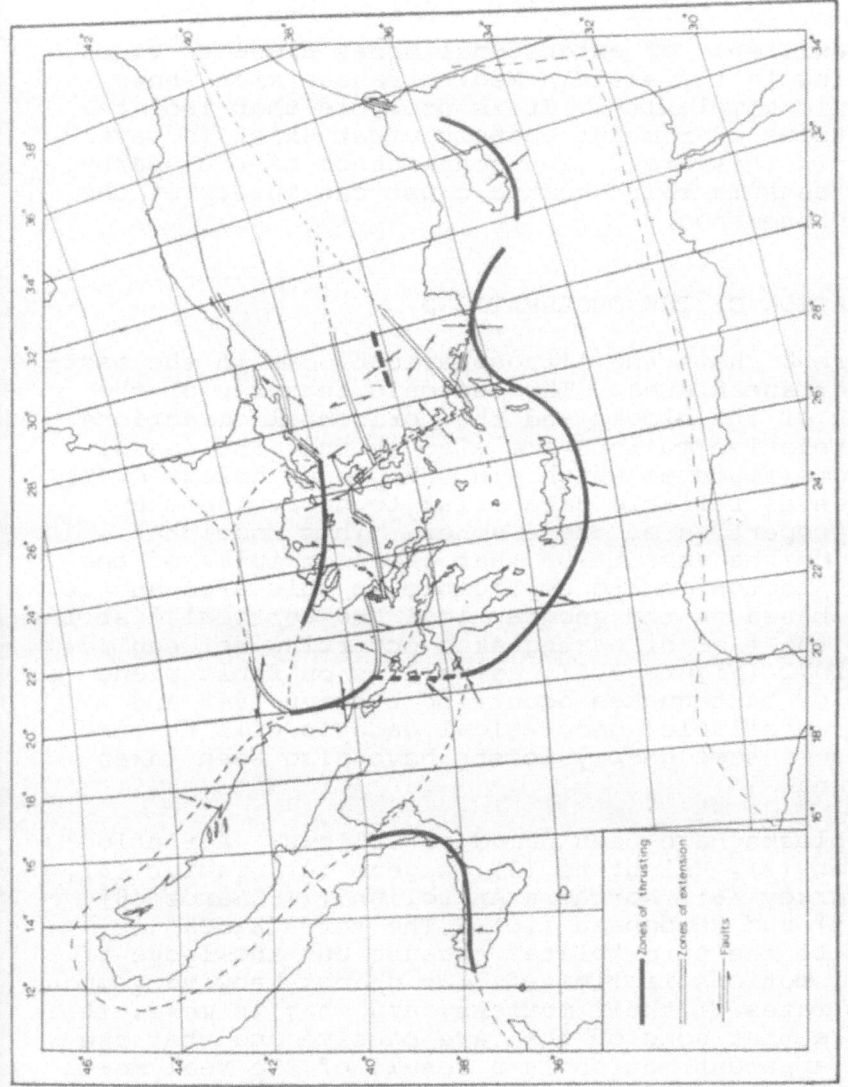

Figure 7: Tectonic Features of Seismic Zones in the
Eastern Mediterranean Area.

(Karnik.[20]). On the other hand, the fault plane solution of one earthquake which occurred on March 11, 1963 (38.1 N, 29.3 E), determined by Canitez and Ucer[29], shows that the slip vector has a considerable thrust component in the proper direction.

The existence of extensional zones close to zones of thrusting in the eastern Mediterranean area shows its tectonic complexity. It is probable that isolated dense portions of remnant oceanic crust exist in several places in this area. The convergence of the nearby blocks to consume this oceanic crust can interpret the observed phenomenon.

INTERPRETATION OF THE OBSERVATIONS

Figure 8 shows the lithospheric blocks in the eastern Mediterranean area. The tectonic features of the boundaries of the blocks and the horizontal directions of their relative motions are also shown. The dashed lines denote features which are unknown or inferred. In these cases no reliable data exist to determine the tectonic properties of these zones. This model has been made on the assupmtion that the principles of the new global tectonics are applicable in this area and it is mainly based on the geographical and vertical distribution of the foci of earthquakes occurring between 1901 and 1972 (Figure 4,5), as well as on fault plane solutions of earthquakes occurring between 1948 and 1969. The available geophysical data as well as certain ideas of some geophysicists have also been taken into account.

The blocks have been named as follows: Adriatic (1), Ionian (2), Valentine (3), Aegean (4), Taurus (5), Western Turkey (6), Northern Anatolian (7), Saros (8), Olympus (9) and Rhodopean (10). The term "block" is preferred to the term "plate" because our knowledge of their real motions is limited. We do not know very much about the rates of their motions, and what is more, it is possible that some of them are passive and that the indicated apparent motion is a result of the real motions of the surrounding blocks.

The distribution of the earthquake foci in the Calabrian arc and in the Tyrrhenian sea and the fault plane solutions of the earthquakes in this area (Ritsema[39]) indicate that the arc is moving in an EES direction. Since seismic and other geophysical information

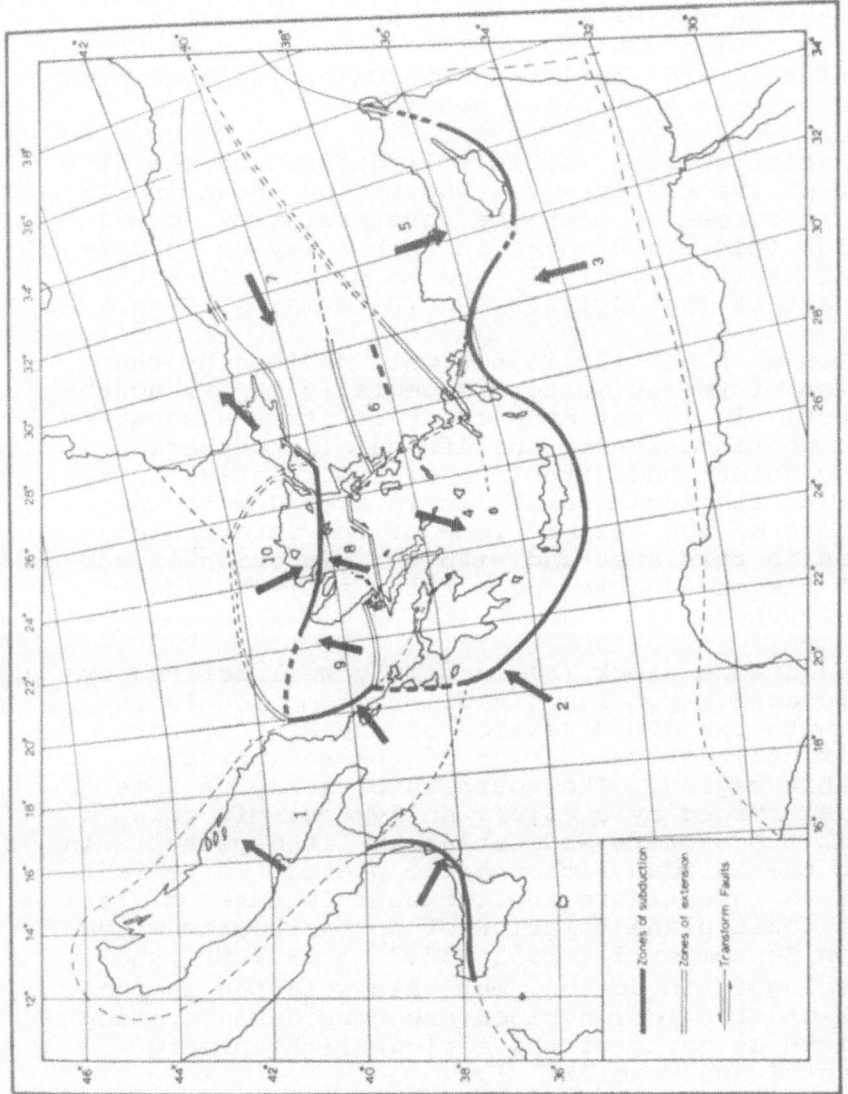

Figure 8: Tectonic Features of Lithospheric Blocks in
the Eastern Mediterranean Area.

shows that this arc is similar, in many respects, to
typical island arcs, such a motion from the concave to
the convex side is in agreement with concepts of the
new global tectonics (Wilson.[61]).Ritsema[5] has presented
evidence that an active sub-thrusting of the lithosphere of the Ionian sea under the Calabrian arc is extremely unlikely and that it is the Calabrian arc which
actively overrides the Ionian sea region.

The Adriatic block (1) is well defined by the distribution of the epicenters. The motion shown in Figure 8 is in agreement with the interpretation according to which this block moves in such a way as to produce normal faulting in Italy and thrusting along the
eastern coast of the Adriatic sea (Ritsema, [39] McKenzie.[32]).

The Ionian block (2) is not well defined by the
distribution of the epicenters especially in its southwestern part. It is not even clear if it is a separate
block or a continuation of the African lithospheric
plate. The determined direction of its relative motion
in respect to the Aegean block is in accord with the
dip direction of the seismic zone of the intermediate
earthquakes in this area and with the direction of subduction of the presumable ocean floor as it is indicated in Figure 3.

The Valentine block (3) is fairly well defined by
the seismic activity. Its direction of motion is in
agreement with the dip direction of the Benioff zone
and the direction of subduction of the Mediterranean
floor in this region. The southern boundary of this
block is determined by a mildly active seismic zone,
along which the eastern Mediterranean lithosphere is in
touch with the African lithospheric plate. No reliable
data exist to investigate the tectonic features of this
zone. The fault plane solution of an earthquake which
occurred on September 12, 1955 (32.3^{O}N, 29.7^{O}E) shows
that thrusting is in action, but this solution is poor.
This block and the Ionian block are considered by many
geophysicists as parts of the African lithospheric
plate (Finetti and Morelli.[43]).

The Aegean block is the most discussed one. Its
western, southern and southeastern boundaries are defined by the seismic zone which follows the Hellenic arc
and dips to the concave side of the arc at a mean angle
of 35 degrees. The length of this dipping Benioff zone
is about 300 km . Its northern and eastern boundaries
are defined by extensional seismic zones. Its motion

relative to Africa has a SSW direction and a probable
rate of about 3 cm/year. For the motion of the Aegean
plate, relative to Europe, McKenzie[32] has estimated a
rate equal to 2 cm/year in a SW direction. The present data give the same rate or a little less, between
Aegean and Europe. Although the motion in Figure 8
interprets many of the seismic data in the Aegean region, it cannot account for the interplate shallow seismic activity and focal mechanism. Figure 4 shows that
there is much shallow seismic activity in zones cross-
ing the plate. As a matter of fact the largest shallow
earthquake occurring in this region since the beginning
of the present century is the big earthquake (M=7.6) of
July 9, 1956 which had its epicenter close to the center of the plate. This earthquake has been followed by
numerous aftershocks (Papazachos et al.[62]). On the
other hand, the focal mechanisms of some earthquakes in
the inner part of the Hellenic arc are not related to
this motion of the Aegean plate in a simple way.

The boundaries of the Taurus block (5) are well
defined by the distribution of the seismic foci. The
direction of motion, determined by fault plane solutions is in good agreement with the direction of dip of
the Benioff zone in this area. The same reasoning as
that used by McKenzie[40] to explain the motion of the
Aegean plate can be applied in this case too. The Taurus block is moving in the direction shown in Figure 8
to consume the Mediterranean sea floor along the thrust
zone which extends at least up to the eastern end of
Cyprus, as it is indicated by the distribution of the
epicenters of the intermediate earthquakes (Figure 5).
According to Ergin (personal communication 1973) there
is geological evidence that this thrust zone extends
further along the boundary between the Turkey and Arabian plates.

The western Turkey block (6) is probably surrounded
by extensional zones. The possible existence of a sinking area in the interior of this block can probably explain this situation. Another possibility is that this
is a passive block and that the properties of its three
boundaries are determined by the outward motion in all
three directions. Further data are needed to obtain
more definite information in respect to the tectonics
of this region.

The determined directions of motions of the (7),
(8), (9) and (10) blocks show that all four blocks converge on the thrust zone which crosses the northernmost

part of the Aegean sea. It is possible that a remnant of old oceanic crust is consumed there by these motions.

We cannot find the direction of thrusting in this zone by the existing data but we can probably estimate the rate of this thrusting. The maximum determined focal depth in this zone is about 100 km . If we assume a length equal to 150 km for the Benioff zone, we shall not be in great error. If the sinking started 10 My ago (Isacks et al[58]), the mean rate of thrusting would be 1.5 cm/year.

There are no fault plane solutions to determine the tectonic properties of the northern boundary of the Rhodopian block (10) but there are some field observations which indicate that extension took place during the occurrence of two large earthquakes in Bulgaria on April 14 and 18 1928 (Richter.[63]).

DISCUSSION

Although the seismotectonic model of Figure 8 interprets well many of the seismic and other geophysical data in this area, it cannot account for certain observations. We have already mentioned that several zones of shallow earthquakes which cross the Aegean plate and the fault plane solutions of these earthquakes are not directly related to the plate motions shown in Figure 8. The occurrence of intermediate earthquakes at considerable depths in the convex side of the Hellenic arc (Papazachos[2]) is another observation which is not easily explained. Ritsema[64] has also shown that certain geophysical properties of the Calabrian arc do not resemble properties of the Pacific island arcs. Such observations show that the tectonics of the eastern Mediterranean area are too complicated to be fully understood by simple seismotectonic models.

The determination of the boundaries and movements of several more lithospheric blocks can probably interpret several of these observations, but by the present accuracy of the focus coordinates and focal mechanism parameters we cannot go much further with respect to this subject. The possibility of the existence of other contributing agents to the complexity of this area, such as flow of material in the mantle, bending of certain tectonic elements and subsidence of crustal blocks, must be seriously taken into account. In addition to the collection of more accurate data on the location and focal mechanism of earthquakes, attempts must be made to determine anomalous zones with respect to the seismic wave attenuation and wave velocity and, in general, to investigate the crustal and upper mantle structure in much more detail than previously. The seismotectonic model of Figure 8 is proposed as a working hypothesis. Only synthetic work based on more accurate seismic as well as on other geophysical and geological data can lead to a satisfactory theory regarding the so complicated active tectonics of the eastern Mediterranean.

REFERENCES

1. Smith, A.G., 1971. Alpine deformation and the oceanic areas of the Tethys, Mediterranean and Atlantic, "Geol.Soc.Am.Bull.", 82, 2039.

2. Papazachos, B.C., 1973. Distribution of seismic foci in the Mediterranean and surrounding area and its tectonic implication, "Geophys. J.Royal Astr. Soc.", (in press).

3. Rabinowitz, P.D., and Ryan, W.B.F., 1970. Gravity anomalis and crustal shortening in the eastern Mediterranean, "Tectonophysics", 5-6, 585.

4. Dewey, J.F., Pitman, W.C., Ryan, W.B.F., and Bonnin, J., 1973. Plate tectonics and the evolution of the Alpine system, "Geol.Soc.Am.Bull.", (in press).

5. Ritsema, A.R., 1970 a. On the origin of the western Mediterranean sea basins, "Tectonophysics", 10, 609.

6. Mallet, R., 1862. Great Neapolitan earthquake of 1857. Chapman and Hall, London, 2 vols.

7. Schmidt, J., Studien über Vulkane und Erdbeben, Leipzig, 1881.

8. Ballore, F. de Montessus de, 1900. La Crecia sismica, "Boll.Soc.Sismol. Italiana", 6, 115.

28

9. Sieberg, A., 1932. Untersuchungen über Erdbeben und Bruchschollenbau im östlichen Mettelmeergebiet, Denkschriften der med.naturw. Ges.zu Jena, 19.Band, 2.Lief.

10. Mihailović, J., 1933. La séismicité de la Thrace, de la Mer de Marmara et de l'Asie Mineure, UGGI, Assoc. Int.de Séism.,Monogr. et Tr. Sc., S.B., F.2.

11. Magnani, M., 1946. Tettonica e sismicita nella regione Albanese, "Geol. Pura e App.", 8, 1.

12. Pinar, N., 1953. Relations entre la tectonique et la séismicité de la Turquie,"Bull.Inf.UGGI",2,261.

13. Ketin, I., and Roesli, F., 1953. Makroseismische Untersuchungen über des nord-westanatolische Beben vom 1953 März 18, "Eclogae Geol.Helv."46,187.

14. Galanopoulos, A., 1965. The large conjugate fault system and associated earthquake in Greece, "Annal. Geol.Pays Hellen.", 18, 119.

15. Galanopoulos, A., 1967. The Seismotectonic regime in Greece, "Ann. Gi Geof.", 20, 109.

16. Galanopoulos, A., 1968. The earthquake activity in the physiographic provinces of the eastern Mediterranean sea, "Annal. Geol. Pays Hellen.", 21, 178.

17. Ergin, K., 1966. On the epicentre map of Turkey and surrounding area, "Turk. Jeol. Kur. Bult.", 10, 122.

18. Gutenberg, B., and Richter, C.F., 1954. Seismicity of the Earth and associated phenomena, 2nd Ed., Princetown Univ. Press. Princeton, N.J.

19. Ergin, K., Guclu, U., and Uz, Z., 1967. A catalogue of earthquakes for Turkey and surrounding area, "Publ. Tech. Univ. Istanbul", 24, 1.

20. Karnik, V., 1969. Seismicity of the European area, part 1, D.Reidel, Dordrecht, Netherlands.

21. Papazachos, B.C., and Comninakis, P.E., 1972. The seismic activity in the area of Greece between 1911 and 1971. Athens, Greece.

22. Karnik, V., 1970. Seismicity of Europe, Progr.Rep. VI, XII Assembl.Gener. Comiss. Seism. Europ. Luxembourg, Sept. 1970.

23. Caputo, M.G., Pamza, F., and Postpischl, D., 1970, Deep structure of the Mediterranean basin,"J. Geophys.Res.", 75, 4919.

24. Papazachos, B.C., and Comninakis, P.E., 1971. Geophysical and tectonic features of the Aegean arc", J.Geophys. Res.", 76. 8517.

25. Hodgson, J., and Cock, J., 1956. Direction of faulting in the Greek earthquakes of August 9-13, 1953. "Publ. Domin. Observ.", 18, 149.

26. Papazachos, B.C., 1961. A contribution to the investigation of earthquake mechanism in Greece, Sci.D.Thesis, Univ. Athens, 75 pp.(in Greek).

27. Ocal, N., 1960. Determination of the Mechanism of some Anatolian earthquakes, "Publ. Dom. Obs. Ottawa", 24, 365.

28. Ocal, N., 1966. Geometrical solutions of fault -plane problem of some of the destructive earthquakes occurring in Anatolia in the period 1938-1955, "Z.Geophys.", 32, 293.

29. Canitez, N., and Ucer, S.B., 1967b. A Catalogue of focal mechanism diagrams for Turkey and adjoining area, ITU Arz. Fizigi Enstitusu.

30. Wickens, J.A., and Hodgson, J.H., 1967. Computer re-evaluation of earthquake mechanism solution 1922-1962. "Publ. Domin. Obs.", 33,1.

31. Delibasis, N., 1968. Focal mechanism of intermediate earthquakes in the area of Greece and their intensity distribution, Sci. D.Thesis, Univ. Athens, 105 pp (in Greek).

32. McKenzie, D.P.,1972. Active tectonics of the Mediterranean region, "Geophys. J.Royal Astr.Soc.", 30, 109.

33. Canitez, N., 1972. Source mechanism and rupture propagation in the Mudurnu valley, Turkey,earthquake of July 22, 1967,"Pure and Appl.Geophys." 93, 116.

30

34. Scheidegger, A.E., 1964. The tectonic stress and
    tectonic motion direction in Europe and western
    Asia as calculated from earthquake fault plane
    solutions, "Bull.Seism. Soc. Am.", 54, 1519.

35. Constantinescu, L., Ruprechtová, L., and Enescu,D.,
    1966. Mediterranean - Alpine earthquake mechan-
    isms and their seismotectonic implications,"Geo-
    phys. J.Royal Astr.Soc.", 10, 347.

36. Canitez, N., and Ucer, S.B., 1967a. Computer De-
    termination for the fault plane solutions in and
    near Anatolia, "Tectonophysics", 4, 235.

37. Shirokova, E.I., 1967. General features in the
    orientation of principal stress in earthquake
    foci in the Mediterranean - Asian seismic belt,
    "Izvestia", 1, 12.

38. Papazachos, B.C., and Delibasis, N., 1969. Tecton-
    ic stress field and seismic faulting in the area
    of Greece, "Tectonophysics", 7, 321.

39. Ritsema, A.R., 1969. Seismotectonic implications
    of a review of European earthquake mechanism,
    "Geologische Rundschau", 59, 37.

40. McKenzie, D.P., 1970. Plate tectonics in the Medi-
    terranean region, "Nature", 226, 239.

41. Lort, J.M., 1971. The tectonics of the eastern Me-
    diterranean," Rev.Geophys. and Space Phys.", 9,
    189.

42. Comninakis, P.E., and Papazachos, B.C., 1972. Sei-
    smicity of the eastern Mediterranean and some
    tectonic features of the Mediterranean ridge,
    "Geol. Soc. A. Bull.", 82, 1093.

43. Finetti, I., and Morelli, C., 1972. Wide scale di-
    gital seismic exploration of the Mediterranean
    sea, "Bollet. Geof. Teor. Appl.", 56, 291.

44. Georgalas, G., 1962. Catalogue of the active vol-
    canoes and solfatara fields of Greece, part 12,
    Int. Assoc. Volcanol., Rome.

45. Ryan, W.B.F., Stanley, D.J., Hersey, J.B., Fahl-
    quist, D.A., and Allan, T.D., 1971. The tect-
    onics and geology of the Mediterranean sea, "The
    Sea", Maxwell, A. (Editor), John Wiley and Sons,
    N.Y. 4, 387.

46. Emery, K.O., Heezen, B.C., and Allan, T.D., 1969.
    Bathymetry of the eastern Mediterranean sea,
    "Deep-Sea Research", 13, 173.

47. Woodside, J., and Bowin, C., 1970. Gravity anom-
    alies and inferred crustal structure in the east-
    ern Mediterranean sea", Bull. Geol. Soc. Am.",
    81, 1107.

48. Harrison, J.C., 1955. An interpretation of gra-
    vity anomalies in the eastern Mediterranean,
    "Phil.Tans. Roy. Soc. Lond.", A248, 283.

49. Papazachos, B.C., and Comninakis, P.E., 1970. Geo-
    physical features of the Greek island arc and of
    the eastern Mediterranean ridge, C.R. Acad. Sci.
    Paris, 16, 74.

50. Vogt, P.R., and Higgs, R.H., 1969. An aeromagnetic
    survey of the eastern Mediterranean sea and its
    interpretation, "Earth and Planetary Sci. Let-
    ters", 5,439.

51. Allan, T.D., and Morelli, C., 1971. A geophysical
    study of the Mediterranean basin, "Bollet.Geof.
    Teor. Appl.", 13,99.

52. Molnar, P., and Oliver, J., 1969. Lateral varia-
    tions of attenuation in the upper mantle and
    discontinuities in the lithosphere, "J.Geophys.
    Res.", 74, 2648.

53. Utsu, T., 1971. Seismological evidence for anomal-
    ous structure of island arcs with special refer-
    ence to the Japanese region, "Rev.Geophys. and
    Space Phys.", 9, 839.

54. Ritsema, A.R., 1972. Deep earthquakes of the Tyrr-
    henian sea, "Geologie en Mijnbouw", 51, 541.

55. McKenzie, D.P., 1969. The relation between fault
    plane solutions for earthquakes and the direct-
    ions of the principal stresses, "Bull.Seism.Soc.
    Am.", 59, 591.

56. Maasha, N., and Molnar, P., 1971. Earthquake fault
    parameters and tectonics in Africa, "J.Geoph.
    Res.", 77, 5731.

57. Sykes, L.R., Fletcher, J.P., and Sbar, M.L., 1973.
    Contemporary stresses, interplate earthquakes,
    and seismic risk associated with high pressure
    fluid injection wells in N.Y.State, Internation-
    al Colloqium on Seismic Effects of Reservoir Im-
    pounding, 26 to 29 March 1973, London.

58. Isacks, B.L., Oliver, J., and Sykes, L.R., 1968.
    Seismology and new global tectonics, "J.Geophys.
    Res.", 73, 5855.

59. Ketin, I., 1948. Über die tektonischmechanischen
    Folgerungen aus den grossen anatolischen Erdbe-
    ben des letzten Dezenniums, "Geol. Rundschau.",
    36, 77.

60. Ambraseys, N.N., and Zatopek, A., 1969. The Mudur-
    nu Valley, West Anatolia, Turkey, earthquake of
    1967 July 22, "Bull. Seism. Soc. Am.", 59, 521.

61. Wilson, T.J., 1965. A new class of faults and
    their bearing on continental drift, "Nature",
    4995, 343.

62. Papazachos, B.C., Delibasis, N., Moumoulidis, G.,
    and Purcaru, G., 1967. Aftershock sequences of
    some large earthquakes in the region of Greece,
    "Ann. Di Geofis.", 20, 1.

63. Richter, C.F., 1958. Elementary Seismology, W.H.
    Freeman, San Fransisco.

64. Ritsema, A.R., 1970 b. Notes on plate tectonics
    and arc movements in the Mediterranean region,

NOTES ON ENGINEERING SEISMOLOGY

by   Professor N.N.Ambraseys,
Imperial College of Science and Technology
London, England.

In some instances the return period of destructive
earthquakes in a particular country is taken as a measure of seismicity. Table 1 shows the return period of
earthquakes that have caused fatalities in 41 countries
during the last 73 years. In this table countries have
been arranged in an increasing order of return period.
Thus earthquakes in Turkey have been taking place 5
times more often than in China. The question is, can
we say that Turkey is 5 times more seismic than China,
or that Iran is 56 times more seismic than Romania? In
some other cases, it is the dimensions of the disaster
that measures the seismicity of a country. Table 2
shows the same 41 countries, this time arranged in a
decreasing order of the total number of fatalities. In
this case China comes first with more than a quarter
million fatalities, five times more "seismic" than Turkey.

It is not difficult to show that it is nearly al.
ways wrong to assume that where damaging earthquakes
happen damage, loss of property and casualties are likely to be high. There are many other parametres that
affect such an idealized relationship; stable population density, degree of concentration of national
wealth, gross national product, and the seismic activity in a country play an important role and this may be
investigated in the following manner.

Countries listed in Tables 1 and 2 were assigned
serial numbers N 1 in increasing order of the return

COUNTRY	RETURN PERIOD OF EARTHQUAKES THAT CAUSED FATALITIES (years)
1. Turkey	0.9
2. Iran	1.3
3. Japan	1.4
4. Greece	1.6
5. U.S.S.R.	2.3
6. Peru	2.5
7. Mexico	2.7
8. Taiwan	2.8
9. Chile	2.9
10. Italy	3.0
11. U.S.A.	3.8
12. China	4.5
13. Albania	4.8
14. Yugoslavia	5.0
15. Pakistan	6.0
16. Algeria	6.1
17. Colombia	6.6
18. Ecuador	6.7
19. India	9.0
20. Argentina	9.1
21. El Salvador	9.2
22. Venezuela	12.0
23. Costa Rica	12.0
23. Afganistan	12.1
24. Burma	18.3
25. Nicaragua	24.0
25. Ethiopia	24.0
25. New Zealand	24.0
25. Morocco	24.0
25. Bulgaria	24.0
25. South Africa	24.0
26. Guatemala	24.3
27. Spain	36.5
27. Portugal	36.5
27. Israel	36.5
27. Libya	36.5
28. Cyprus	73.0
28. Lebanon	73.0
28. Iraq	73.0
28. Tunisia	73.0
28. Romania	73.0

Table 1: Return period of earthquakes that caused fatalities

COUNTRY	TOTAL NUMBER OF PEOPLE KILLED	POPULATION DENSITY km^2
1. China	269,279	75
2. Japan	163,245	268
3. Italy	92,465	174
4. U.S.S.R.	67,089	11
5. Iran	58,176	16
6. Peru	57,308	10
7. Turkey	56,799	40
8. Pakistan	54,448	113
9. Chile	37,827	12
10. Nicaragua	13,502	12
11. Morocco	11,511	29
12. India	8,523	156
13. Taiwan	6,969	357
14. Argentina	6,228	8
15. Ecuador	5,864	12
16. El Salvador	2,989	148
17. Costa Rica	1,726	31
18. Colombia	1,587	17
19. Greece	1,563	66
20. Algeria	1,499	5
21. Yugoslavia	1,326	78
22. Mexico	1,302	24
23. Burma	831	38
24. U.S.A.	791	22
25. Venezuela	577	10
26. Albania	551	70
27. Afganistan	534	24
28. Romania	469	102
29. Libya	450	1
30. New Zealand	276	10
31. Israel	264	130
32. Bulgaria	252	75
33. Lebanon	136	284
34. Ethiopia	85	23
35. Cyprus	40	67
36. Guatemala	43	44
37. Portugal	22	102
38. South Africa	17	85
39. Tunisia	14	27
40. Spain	8	64
41. Iraq	6	19

Table 2: Seismicity indicated by ordered, total number of fatalities

36

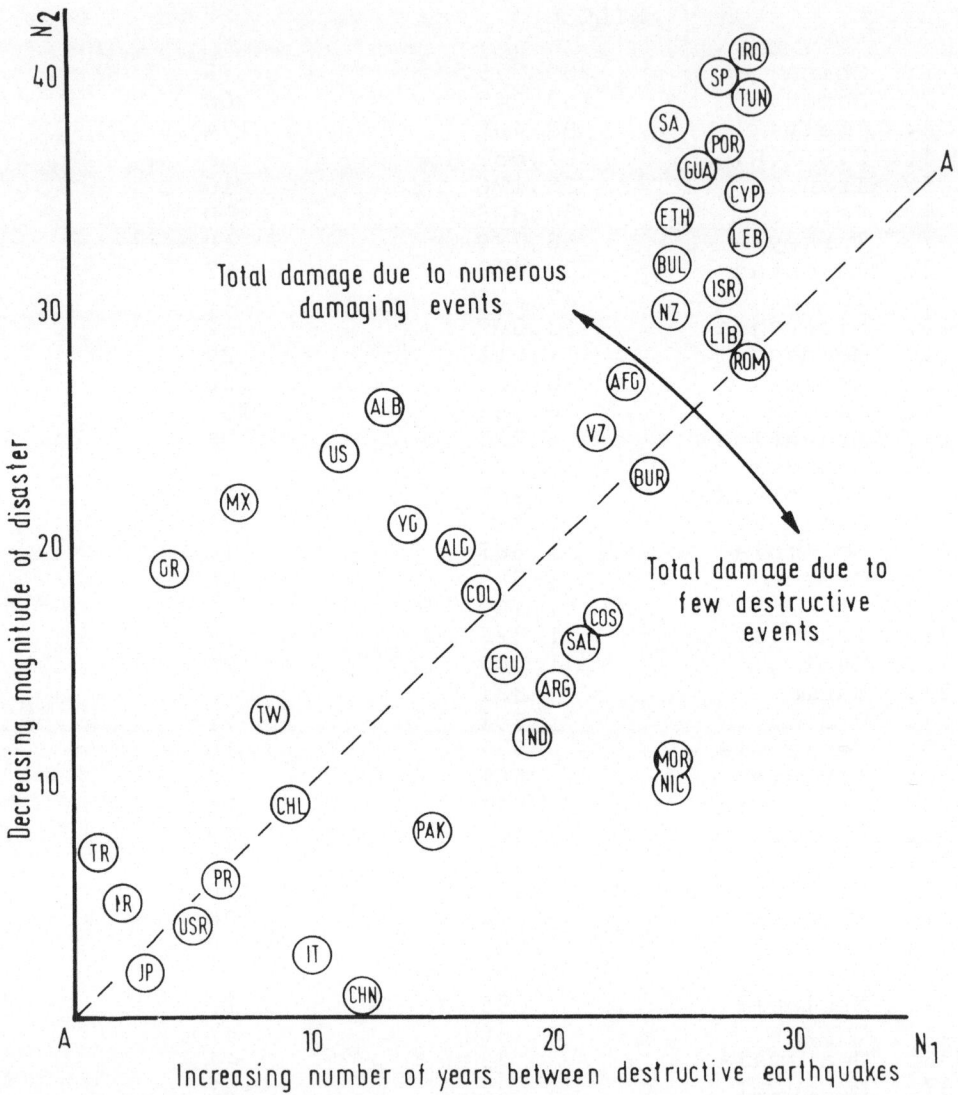

Figure 1: Scatter diagram of results from Tables 1 and 2 .

period (Table 1), and N 2 in decreasing order of fatalities (Table 2). A scatter plot, Figure 1, is then
constructed, in which the abscissa of each point is the
serial number N 2 , while the ordinate is the corresponding serial number of the country in the return period list N 1 .

On the scatter plot, an ideal relationship between
relative return periods and casualties should lead to
concentration near the diagonal A-A . Clustering away
from the diagonal towards the N 2-axis indicates
total damage during the last 73 years caused by numerous events, while clustering towards the N 1-axis
suggests damage due to few earthquakes.

Like other semi-empirical measures, intensity,
when assessed at a large number of points, may show
regular distribution patterns which, together with We-
ber-Fechner's law of Intensity, have helped seismologists to determine relative focal depths, energy absorption coefficients and other less earthquake parametres.

There is, however, one common feature which belongs to Intensity, namely the use of its variation
from place to place, rather than its absolute value.
Intensity, being a convenient means of conveying in a
single rating of the Scale a measure of the effect of
ground motion on man-made structures, has been adopted
by the engineer for design purposes with little hesit-
ation and perhaps with even less understanding of the
subtleties which are involved in its definition and
assessment. The development, by the engineer, of Intensity into a useful design tool, obviously had its
roots in more or less practical requirements. The result today is that we have a multitude of empirical
formulae from which the engineer thinks he can confid-
ently derive magnitudes, focal depths, ground accelerations and velocities as a function of Intensity (Table
3).

There seems to be a great danger in the prevailing
over-emphasis on Intensity and on empirical relations
between Intensity and maximum ground acceleration or
velocity. Modern textbooks and current Codes feature
tables and formulae for the conversion of Intensity
into maximum acceleration without saying how these formulae have been obtained and with what scatter of the
data.

Figure 2 shows a plot of recorded maximum ground

$$M = A(I_o) + B(\log h) + C(\log r) + D$$

Region	A	B	C	D	Ref.
Bulgaria	0.50	0	0	+1.10	1
California, South	0.60	+1.80	0	−1.10	2
California, South	0.60	0	0	+1.30	2
California	0.67	0	0	+1.00	2
California	0.62	0	0	+1.14	3
Caucasus	0.69	0	0	+0.90	4
Caucasus	0.93	+1.14	0	−3.00	5
China	0.58	0	0	+1.50	6
China	0.67	+0.80	0	−0.50	7
Czechoslovakia	0.55	+0.93	0	+0.14	8
Europe	0.50	+1.00	0	+0.35	9
Europe, Eastern	0.67	+2.30	0	−2.00	10
Germany, S.Jura	0.80	0	0	−0.90	11
Germany	0.50	+1.00	0	+1.32	12
Greece	0.28	0	1.66	−0.13	13
Greece	1.79	0	3.58	−3.97	13
Hungary	0.60	0	0	+0.30	14
Hungary	0.60	+1.80	0	−1.30	14
Israel	0.50	0	0	+1.80	15
Italy	0.48	0	0	+1.41	16
Italy, Alps	0.69	+2.48	0	−2.06	17
Italy, Calabria	0.69	+3.51	0	−4.31	17
Italy, N.Sicily, Adriatic	0.69	+2.96	0	−2.96	17
Oregon & Washington	0.69	0	0	+0.82	3
Spain	0.42	+1.07	0	+1.49	18
Tadzhikistan, USSR	0.61	+0.67	0	+0.39	19

$$I = I_o - 0.5 - 3.17 \log(D) \qquad\qquad 20$$

$$I = I_7 - 9.66 - 0.0037(D) + 1.38(M) + 0.00528(DxM) \qquad 21$$

$$I_o = 1.5(M-1.1) - 3.5 \log(h - 0.1 \times 10^{o.3M}) + 3.00 \qquad 22$$

$$I_j = 2.0(M) - 9.7 + 2.0 \log(100/R) + 0.0166(R-100) \qquad 23$$

Tabulated and graphical relationships:

I_o, I, R, and h $\qquad\qquad\qquad\qquad\qquad$ 24

I_j, D, M and h $\qquad\qquad\qquad\qquad\qquad\quad$ 25

Table 3: Empirical relationship between Intensity and other parameters.

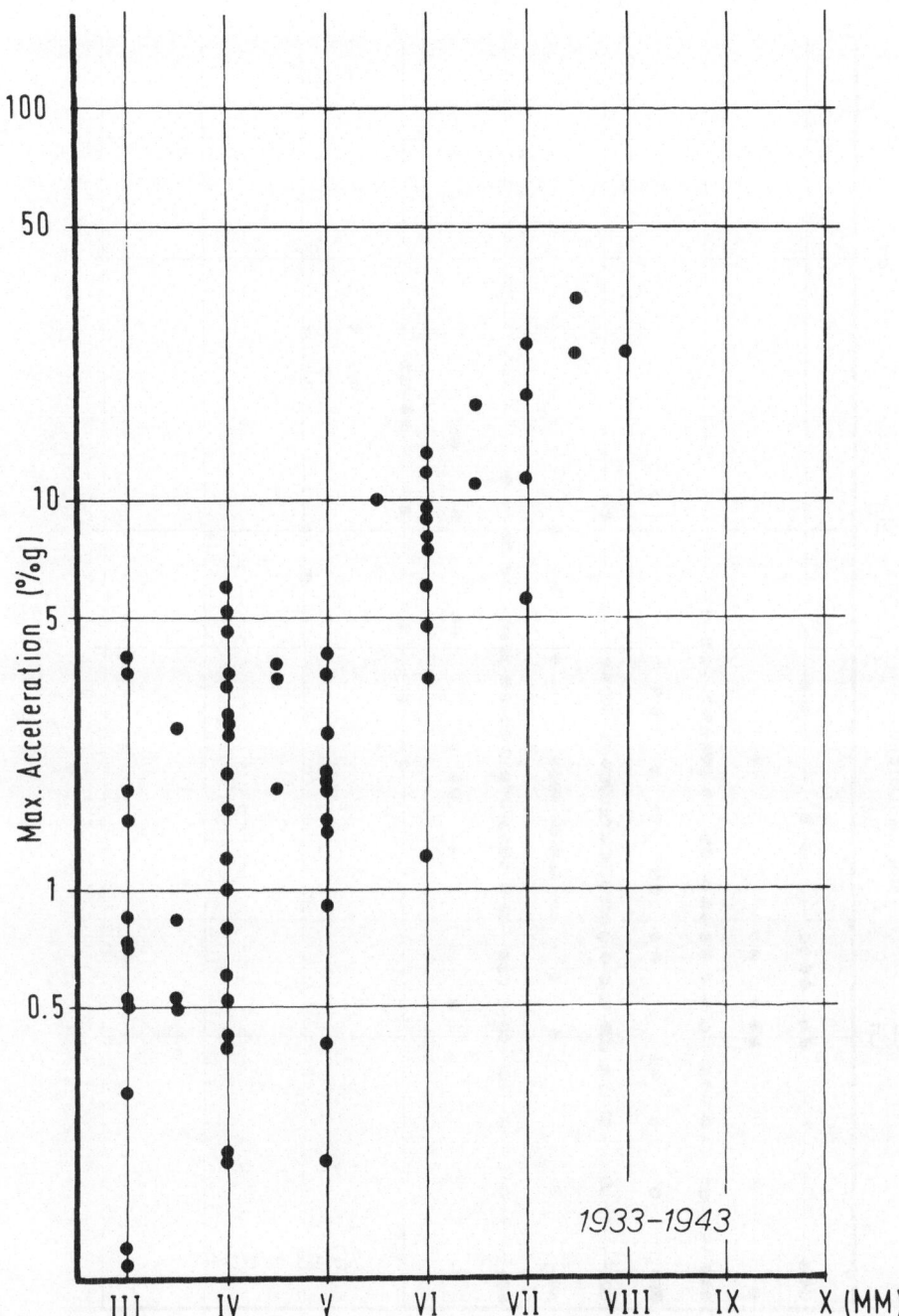

Figure 2: Recorded maximum accelerations VS reported
          Intensities for the period 1933-1943.

40

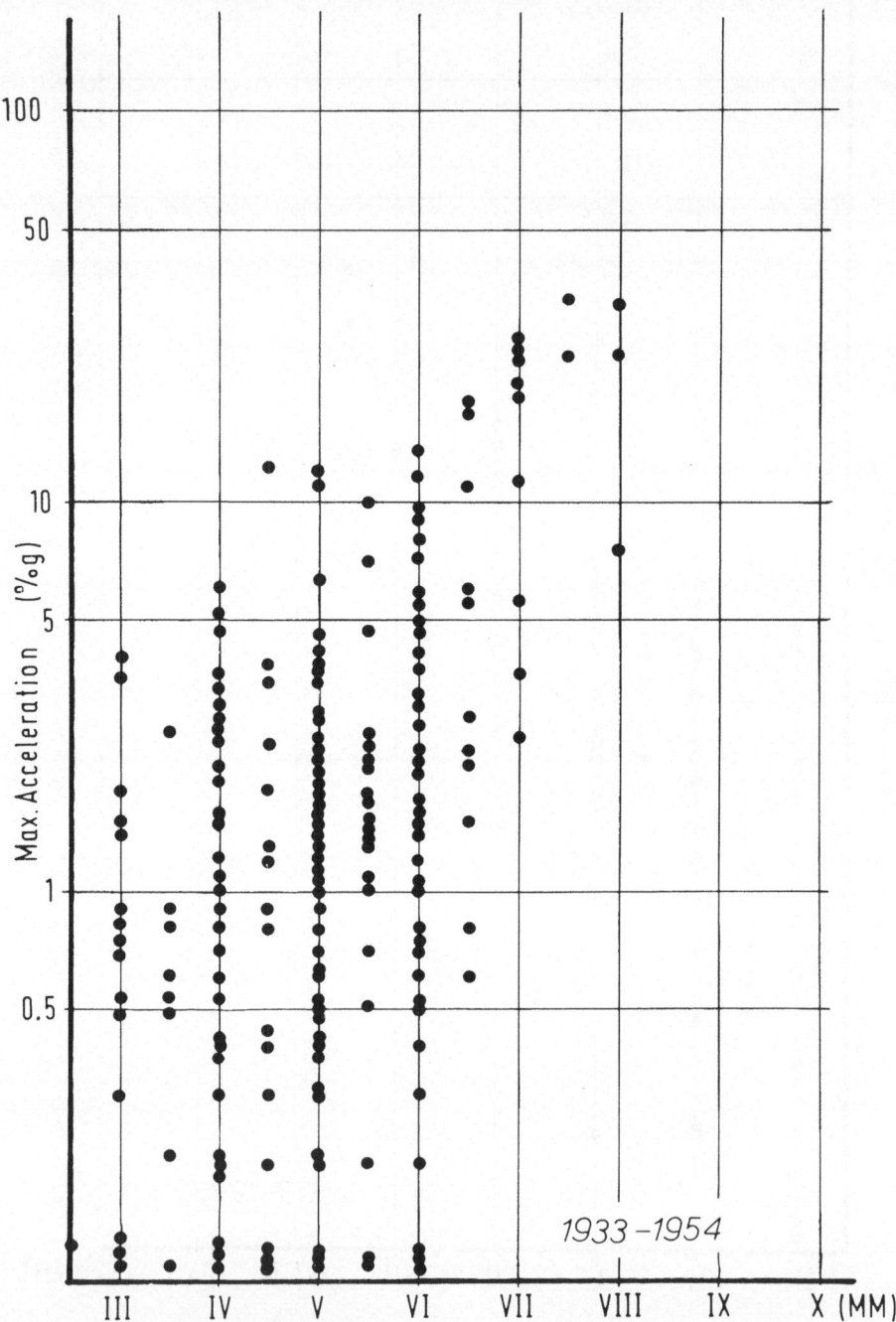

Figure 3: Recorded maximum accelerations VS reported
Intensities for the period 1933-1954.

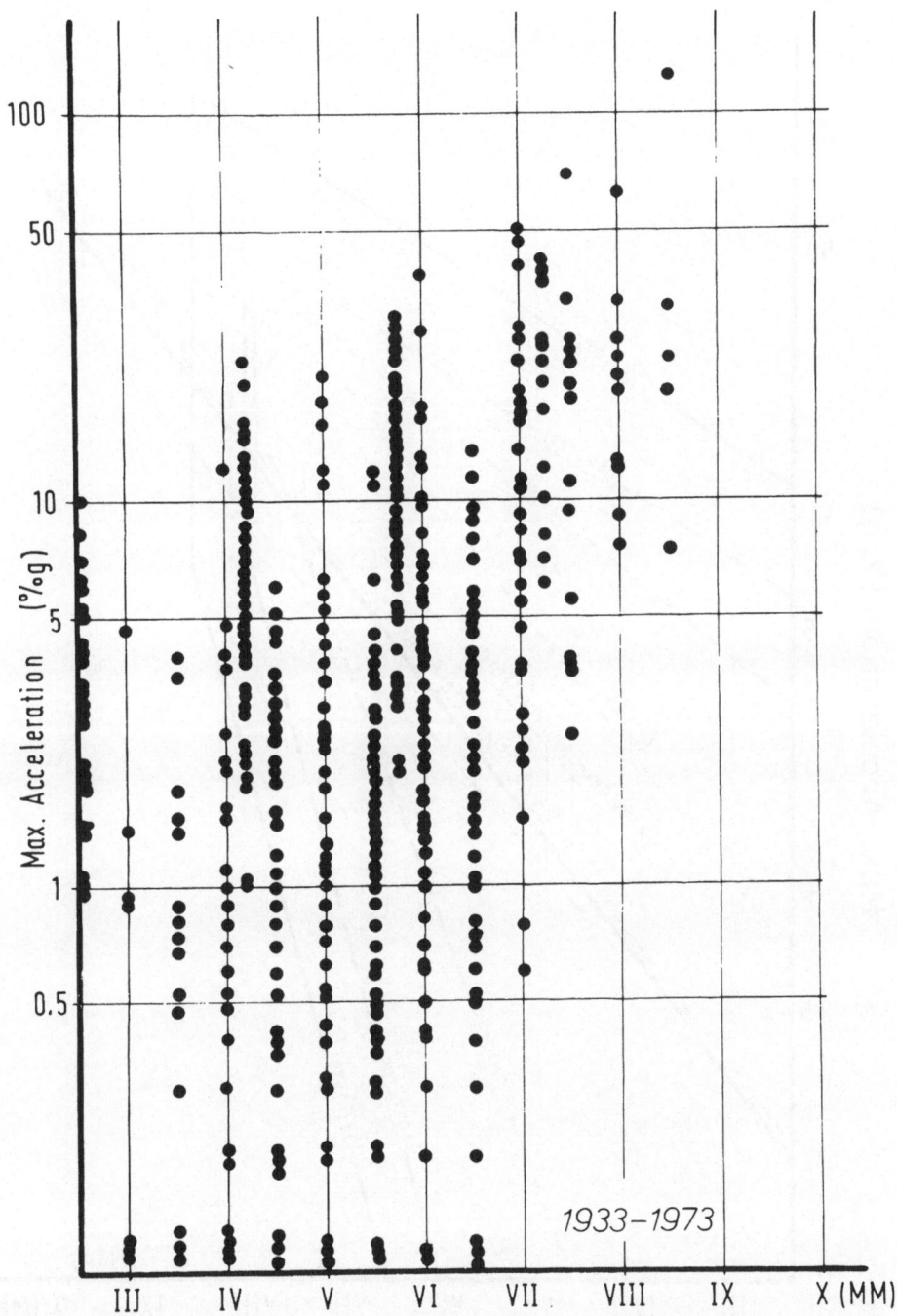

Figure 4: Recorded maximum accelerations VS reported
Intensities for the period 1933-1973.

42

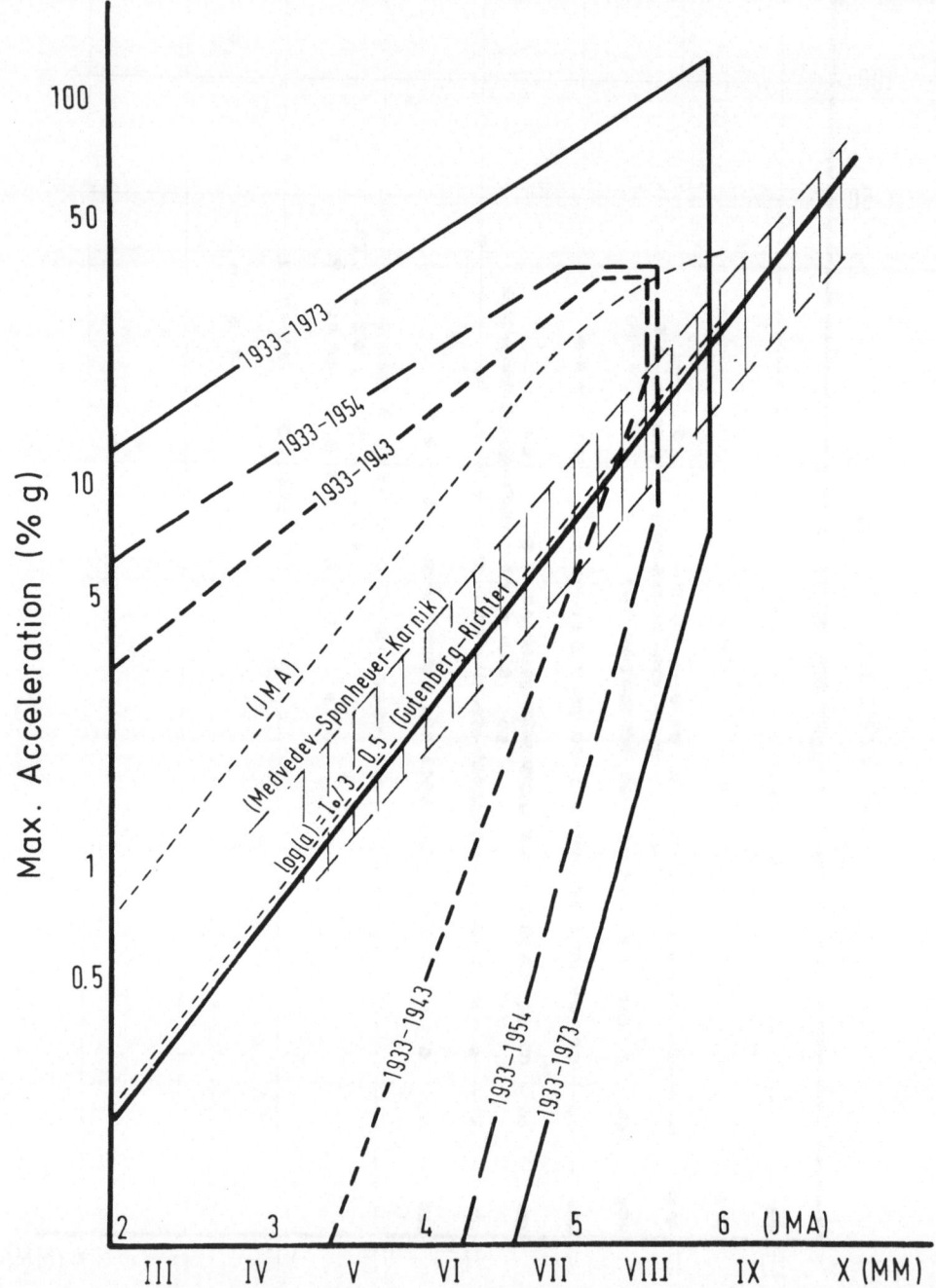

Figure 5: Development of acceleration-Intensity relationships.

accelerations versus reported Intensities at the recording station for the period 1933-1943. The scatter
is considerable; nevertheless there is an obvious
correlation trend which, in 1943, should have tempted
one to fit a curve through these points and, from this
11-year sample, obtain an empirical formula. Figure 3
shows a similar plot for a 20-year period and fitting a
curve to this plot should have required considerable
courage. Today, fitting a curve to such a plot becomes
impossible, Figure 4. The scatter is too great to allow the assumption of any possible correlation between
Intensity and acceleration. Figure 5 seems adequate to
overcome one's initial doubts that such relationships
could not exist. This, however, by no means implies
that Intensity should be abandoned as a means of conveying useful information of earthquake effects in a
coded form. In engineering seismology Intensity can be
as useful as Atterberg's limits are in Soil Mechanics.

It must be understood, however, that Intensity, as
defined by any of the current Scales, cannot be assessed
rigorously, nor when assessed can it be used for design
purposes. It is true that there is great difficulty in
abandoning established usages, but one should think
that there is more need for instrumental data from
simple and perhaps cheap instruments such as seismoscopes, and fewer attempts at inventing new Intensity
Scales, particularly "international" ones, or at prod-
using new empirical formulae.

ENGINEERING EVALUATION OF MAXIMUM GROUND ACCELERATIONS

One of the most popular parametres in the earthquake resistant design of structures is the ground acceleration, and its prediction has attracted the attention of many seismologists and engineers. In the last
few decades numerous attempts have been made to correlate maximum ground acceleration with epicentral or focal distance, with Intensity, Magnitude and with other
parametres that describe the earthquake source and medium through which seismic energy has to travel before
it reaches the ground surface. Table 4 shows some of
the results of these attempts, most of which are based
on a pseudo-statistical treatment of a body of heterog-
eneous data.

From the study of 70 strong-motion records of vect-
orially resolved acceleration equal to or greater than
20% g, we find that for all practical purposes there is

$$\log(a) = 0.331(I) - 0.923 \qquad 26/27$$
$$= 0.416(I) - 1.040 \qquad 28/29$$
$$= 0.330(I) - 0.500 \qquad 2/30$$
$$= 0.500(I) - 1.350 \qquad 11$$
$$= 0.427(I) - 0.897 \qquad 31$$
$$= 0.550(I) - 1.222 \qquad 25$$
$$\log(\bar{a}) = 0.308(I) - 0.040 \qquad 20$$
$$\log(a) = 0.500(I_j) - 0.347 \qquad 23$$
$$\log(a) = 4.79 - 1.92 \log(D) \qquad 20$$
$$\log(a) = 3.31 + 0.34(M) - 2.0 \log(R) \qquad 32$$
$$\log(a) = 3.09 + 0.347(M) - 2.0 \log(R+25) \qquad 33$$
$$\log(a) = -1.75 + 0.466(M) - 1.4 \log(R) \qquad 34$$
$$\log(a_o) = -2.1 + 0.81(M) - 0.027(M)^2 \qquad 2$$
$$\log(a) = 0.698 - 0.5 \log(T) + 0.61(M) -$$
$$(1.66 + 3.6/R) \log(R) + (0.167 - 1.83/R) \qquad 35$$
$$\log(a/g) = -0.2 + 0.20(M) - 1.10 \log(R) \qquad 36$$
$$\log(a/g) = 3.0 - 2.0 \log(1.6 D + 43) \qquad 37$$
$$\log(a/g) = 3.5 - 2.0 \log(1.6 D + 80) \qquad 37$$
$$\log(a/g) = \log(0.69 x e^{1.6M}) - \log(1.1 o x e^{1.1M} + D^2) \qquad 21$$
$$a/g = 335.5/((1.6 D)^2 + 225)^{-1} \qquad 38$$

Tabulated and graphical relationships: I, a   39

M, I_o, a_o, r   40

M, D, a   40

M, D, a   41

I, R, a   42

I, R, a   43

Table 4: Empirical relationships between acceleration
and other parameters

no significant correlation between Magnitude, distance
and acceleration in the near-field. At large distances
or for small accelerations these three variables become
more interdependent, but then such low accelerations
are of little interest to the designer.

It has always been recognized by engineers that in
the epicentral region of a strong earthquake ground accelerations are large. Before the advent of strong
-motion recorders some idea of their magnitude was
gleaned indirectly from the overturning of simple objects. The following pre-1949 earthquakes showed comparatively large accelerations:

Neapolitan	1857	...	30% g	Messina	1908	...	20% g
Achaia	1862	...	37% g	Tango	1927	...	47% g
Mino-Oward	1891	...	43% g	Ogasima	1934	...	73% g
Assam	1897	...	42% g	Nankai	1946	...	47% g
San Fransisco	1906	...	20% g	Imaichi	1949	...	95% g

In some of these events the apparent acceleration
exceeded 100% g. In 1933 reliable accelerations became
available with the first good strong-motion record from
the Long Beach earthquake. This record showed an un-
disputable 23% g , which vindicated those who doubted
the much larger, pre-1933, accelerations.

A few years later, however, in 1940, the El Centro
record displayed even larger accelerations of over
30% g. The question then arose as to whether these were
typical of a strong earthquake and whether accelerations
of this magnitude should in practice be used. For a
few years there was no consensus of opinion among engineers about the suitability of this record for design
purposes. Up until that time observations over a whole
century had failed to produce conclusive evidence for
accelerations much larger than those recorded at El
Centro, Figure 6. Therefore, engineers began to wonder
if ground accelerations did in fact have an upper bound,
and values for this limit were put forward, that of
50% g gaining considerable popularity among designers.

The question of the 50% g upper bound was, however,
soon resolved. A series of strong-motion records from
Matsushiro, Parkfield, Koyna, Pacoima and Stone Canyon,
show not only that there is no such upper bound, but
also that acceleration per se is not the most significant design parametre. The maximum values from these
records of 52% g, 51% g, 63% g, 124% g and 70% g respectively were received by designers with mixed feel-

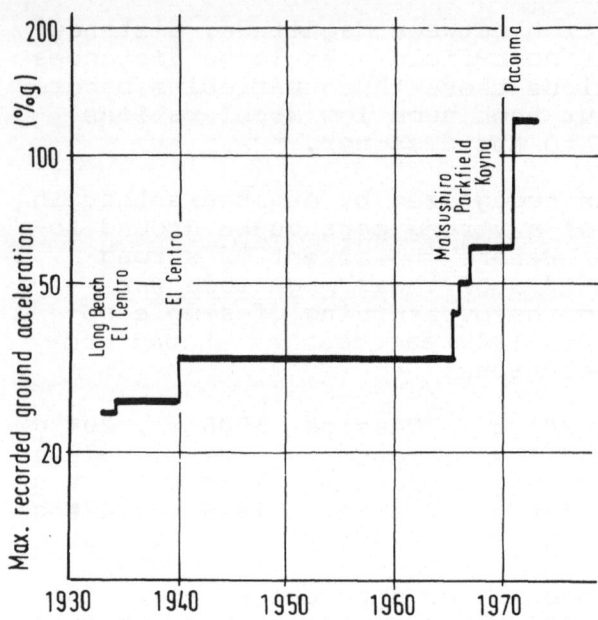

Figure 6a: Maximum recorded ground accelerations ever.

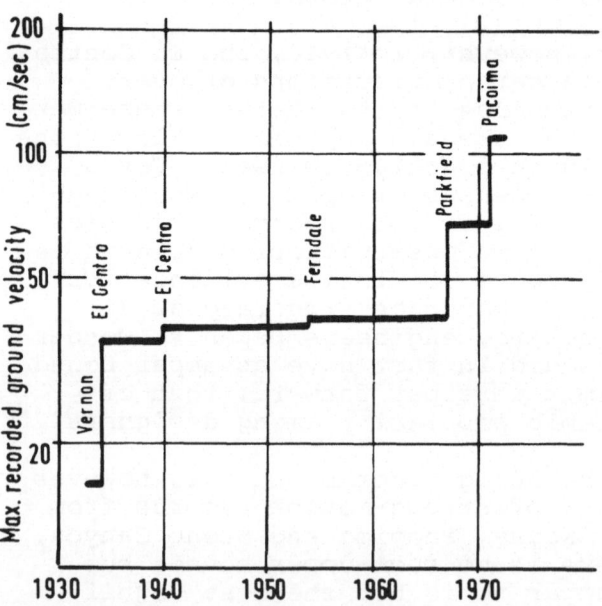

Figure 6b: Maximum recorded ground velocities ever.

ings. Dissentients argued, and still do, that these records are not representative of the free-field motion, and that they have been influenced by local site and topographic conditions, arguments reminiscent of those in 1941 when the El Centro record was under judgment. But in one way or another all strong-motion recordings are influenced by local site conditions, in the same way that an engineering structure would be if it was located at the site of the instrument.

The question of whether there is an upper bound for ground accelerations is indeed of great importance to the engineer. It cannot be answered, however, by treating the ground as a purely elastic, hysteretic medium. This is because the material through which acceleration pulses have to travel before reaching the surface, be it alluvium rock, has a finite shear strength, and stress waves under certain conditions may bring about internal or surface failure, with the result that accelerations above a certain value will be prevented from reaching the surface. For competent bedrock, within the earthquake source, analysis predicts maximum accelerations which, depending on the strength-drop during faulting or slip, may reach values well in excess of 200% g.

It can be shown that maximum possible accelerations in homogenous, normally consolidated deposits cannot exceed values of the order of $(c_u/p')(p'/p)g$; in saturated or dry sands these values are $(p'/p)g \cdot \tan\emptyset'$ and $g \cdot \tan\emptyset'$ respectively, upper bounds dictated solely by the properties of the deposit. The implication of these simple relations is obvious. Suppose, for the moment, the effect of cyclic loading on c_u is ignored. Then, a low plasticity, normally consolidated clay should be incapable of transmitting to the surface accelerations greater than 10 to 15% g. On the other hand, consideration of the effects of cyclic loading on c_u wil obviously lead to even lower accelerations, though to large displacements. High plasticity deposits will show 25 to 35% g and saturated sandy clays and medium dense sands may allow as much as 50 to 60% g to go through. In clean gravel and dry dense sands accelerations may reach much higher values.

For strong foundation materials or for small amplitude accelerations the deposit will behave elastically and it may therefore amplify the bedrock motion. As the strength of the material decreases, or as the

$$\log(v) = 0.30(I) - 1.156 \qquad\qquad 44$$

$$\log(v) = 1.204 + 0.43(M) - 1.70 \log(R) \qquad\qquad 33$$

$$\log(v) = 1.176 + 0.43(M) - 1.70 \log(R + 0.17xe^{0.59M}) \qquad 34$$

$$\log(v) = -2.17 + 0.055(M) - 1.5 \log(R) \qquad\qquad 35$$

$$\log(v_m) = 0.31(I) - 0.611 \qquad\qquad 18$$

$$\log(V) = -0.98 + 0.72(M) - 0.5 \log(11.5M - 53.0) - \log(R_o) \quad 45$$

NOTE: a = maximum ground acceleration in cm/sec^2

a_o = maximum epicentral ground acceleration in cm/sec^2

$\bar{a}$ = average maximum acceleration in cm/sec^2

D = epicentral distance in km

I = local intensity (MM) or (MSK)

I_o = epicentral intensity

I_j = local intensity (Jap. Met. Office)

I_7 = intensity of a Magnitude 7 earthquake at the same location

h = focal depth in km

M = magnitude

R = focal distance in km

R_o = focal distance assuming 2h, in km ($R_o \leq 60$ km) (

T = predominant period of ground motion in secs.

v = ground velocity in cm/sec

vm = spectral intensity

V = maximum resultant velocity in cm/sec (3-components)

r = radius of perceptibility in km

(For limitations and range of application of formulae see Refs.)

Table 5: Empirical relationships between velocity and other parameters

intensity of motion increases, the response may bring
about internal or near-surface yielding, as a result of
which accelerations above a certain level will be prevented from reaching the surface.

Clearly these observations raise an obvious
question, to which no obvious answer can be given: why
do Earthquake Codes indiscriminately recommend a base
-shear coefficient which increases with decreasing
strength of foundation materials? Perhaps the authors
of these Codes confuse damage due to foundation failures
with that caused by shaking, two distinctly different
sources of damage that should be treated separately.
However, with regard to seismic problems the ground
acceleration is not the most significant parameter;
rather, it is the amplitude of the maximum ground velocities and to some extent the duration of shaking or
the rate at which energy flux is supplied to a structure. As with accelerations, so with velocities, numerous attempts have been made to correlate them with
Magnitude and focal distance. Table 5 shows some of
the better known relationships which are in principle
open to the same criticism as those derived for accelerations. Near the source, within a distance of one to
two focal depths from the causative fault or focus, attenuation is erratic and shows a weak correlation with
Magnitude and distance, contrary to that expected from
reasonably representative analytical models.

It can be shown that the initial particle velocity
on a fault surface may be expressed by $v = PS(1-p_0)/2G$,
where P and $p_0 P$ are the peak and residual shear
strength on the fault respectively; S is the shear
wave velocity and G the shear modulus of the medium.
Depending on the strength-drop, particle velocities
very near the source may reach values as high as 480cm/
sec. However, for shallow earthquakes laboratory evidence suggests that p_0 is not zero, which, together
with the fact that there has been neither observable
melting on fault planes, nor heat-flow anomalies along
re-activated faults, suggests that the upper bound for
bedrock velocities near a fault break should be about
150 cm/sec.

A simple method for the assessment of permanent
displacements has been developed. It is based on two
parameters; one is k_c , the minimum seismic coeffic-
ent required to bring about incipient failure in an
earth dam or foundation, and the other k_m , the maxi-
mim local seismic coefficient or maximum input acceler-

50

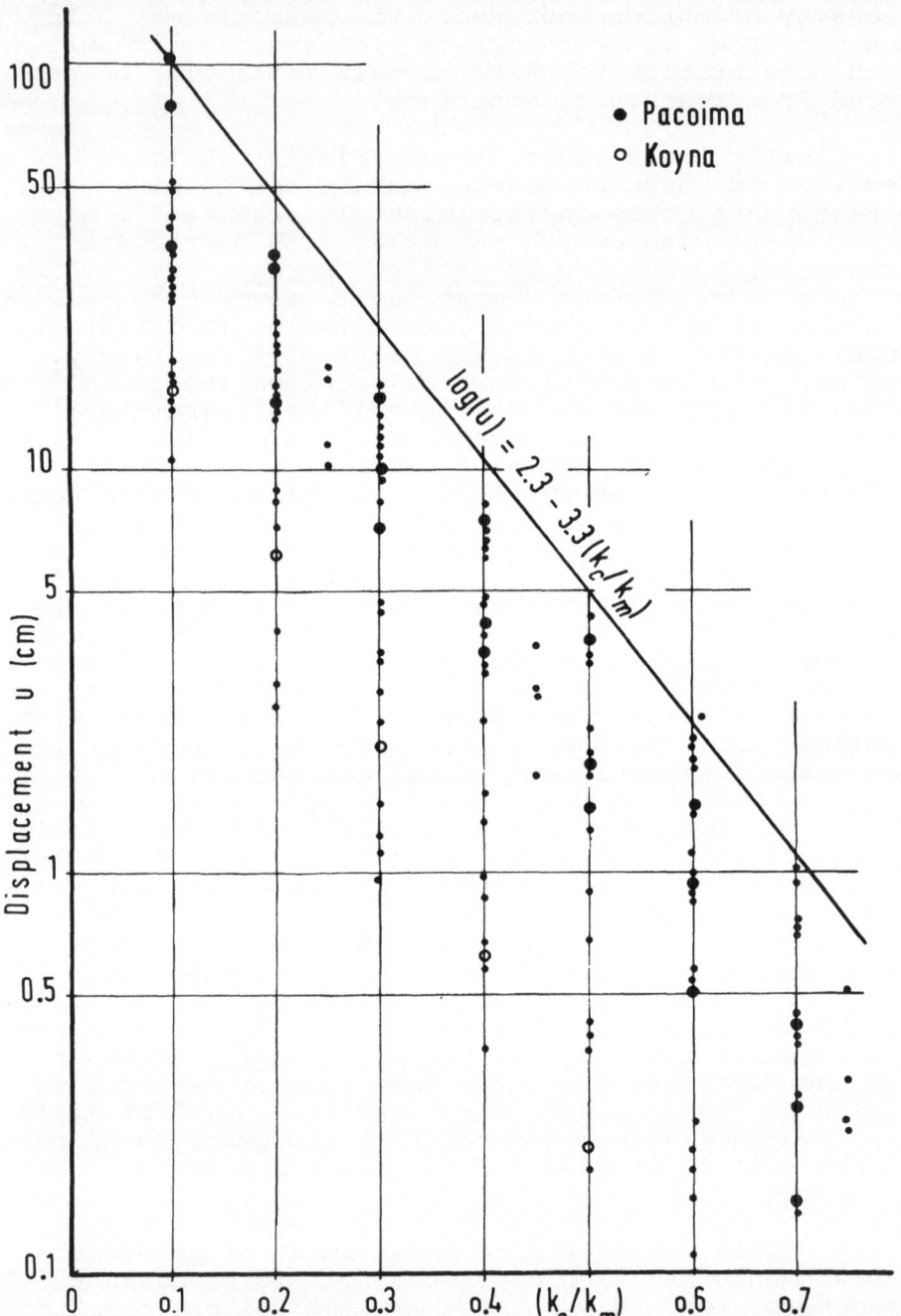

Figure 7: Upper bound of permanent ground displacements for a mass of critical input accelerations k_c.

ation. k_c is a function of the geometry and soil parameters of the sliding mass corresponding to a factor of safety of one. Calculations for a variety of inputs including those for Pacoima and Koyna, as well as from a number of recent high yield under-ground explosions, Figure 7, show that the upper bound of the displacement u for a mass of critical acceleration k_c will be of the order of

$$\log(u) = 2.3-3.3(k_c/k_m) .$$

The displacement u in centimetres is measured along a plane forming an angle b with the horizontal.

REFERENCES

1. Grigorova, E., Grigorov B., Determination des magnitudes des seismes proches d'apres les observations des apareils, Izvest. na Geofiz. Inst.,3, 193, Sofia 1962.

2. Gutenberg B., Richter C., Earthquake magnitude, intensity energy and acceleration, Bull. Seism.Soc. America, 46, 128, 1956.

3. Algermissen S., Stepp J., Rinehart W., Arnold E., Studies in seismicity and earthquake damage statistics, Environm. Sci. Serv. Adm. Report, U.S.Dept. Comm., 2, 1969.

4. Savarenski E., Dzibladze E., On the seismicity of the Great Caucasus, Izvest. Akad. Nauk, Ser.Geofiz., 577, Moscow 1956.

5. Aivazov, I. Zavisimost mezhdu ballnostyu intensivnostyu i glubinoy ochaga dlya kavkazshikh zemletresenii, Subsh. Akad. Nauk Gruz. CCP,27,149,1961.

6. Lee, S., A practical magnitude scale, Acta Geophysica Sinica, 9, 98, 1958.

7. Mei-Shih, Y. O seismicheskoy aktivnosti Kitaya, Izvest. Akadem. Nauk. Ser. Geofiz., 3, 381, Moscow 1960.

8. Karnik, V., Michal E., Molnar A., Erdbebenkatalog der Tschechoslowakie bis zum Jahre 1956, Travaux Instit. Geophys. Acad. Chechosl. Sci., 69, 411, 1957.

9.   Karnik, V., Seismicity of the European area, Publ. Akad. Chechoslovak. Sci., 1, 1968

10. Shebalin, N., Correlation between earthquake magnitude and intensity, Studia Geophys. et Geodet., 2, 86, Praha 1958.

11. Peterschmitt, E., Etude de la magnitude des seismes, Annales Inst. Phys. Globe, New Ser., 6, part Geoph., 51, Strassbourg 1950.

12. Adlung A., Intensitätsbestimmung seismischer Erschütterungen durch einfache Instrumente und deren Beziehung zur Magnitude, Travaux Inst. Geoph. Acad. Tchecosl. Sci., 155, 311, 1961.

13. Galanopoulos, A., On magnitude determinations using macroseismic data, Annali di Geofizica, 14, 115, 1961.

14. Csomor, D., Kiss Z., Die Seismizität von Ungarn, Studia Geophys. et Geodet., 3, 33, Praha 1959.

15. Arieh, E., Seismicity of Israel and adjacent areas, Bull. Geol. Surv. Israel, 43, 1, 1967.

16. Marcelli L., Montecchi A., Contributi per uno studio sulla seismicita dell'Italia, Annali di Geofisica, 15, 159, 1962.

17. Shenkareva, G., Seismicity of Italy, Bollet. Geofisica Teoret. Applic., 13, 271, Udine 1971.

18. Munuera J., El mapa de zonas sismicas generalizadas de la Peninsula Iberica, Publ. Inst. Geograph. y Catastral, 1969.

19. Bune, V., O klasifikatsii zemletresenii po ikh sile no osnove instrumentalniykh nablyudenii, Izvestia Akad. Nauk Ser. Geofiz., 1, 48, 1956.

20. Neumann, F., Earthquake intensity and related ground motion, Univ. of Washington Press, Seattle, 1954.

21. Milne, W., Davenport A., Distribution of earthquake risk in Canada, Bull. Seism. Soc. America, 59, 729, 1969.

22. Shebalin, N., Maximum magnitude and maximum scale intensity of an earthquake, Fizika Zemli, Izvest. Akad. Nauk, 12, 1971.

23. Kawasumi, H., Measure of earthquake danger and expectancy of maximum intensity throughout Japan, Bull. Earthq. Res. Inst., 29, part.3, 472, 1951.

24. Ergin, K., Observed intensity-epicentral distance relations in earthquakes, Bull. Seism. Soc.America, 59, 1227, 1969.

25. Hirono T., Hisamoto S., Amplifying effects of ground on seismic acceleration, Collected Papers Kenzo Sassa, Geoph. Inst. Univ. Kyoto, 69, 1963.

26. Cancani, A., Sur l'emploi d'une double echelle sismique des intensites empirique et absolue, Proc. Intern. Seism. Assoc., Manchester, 1904.

27. Kövesligethy, R., Seismonomia, Modena, 76, 1900.

28. Peterschmitt, E., Sur la variation de l'Intensite macroseismique avec la distance epicentrale, Travaux Sci. Bureau Centr. Intern. Seismol., Ser. A., 18, 183, Strassbourg 1952.

29. Ishimoto, M., Echelle d'intensite sismique et acceleration maxima, Bull. Earthq. Res. Inst., 10, 614, 1932.

30. Gutenber, B., Richter C., Earthquake magnitude, intensity, energy and acceleration, Bull. Seism. Soc. America, 38, 163, 1942.

31. Hershberger, J., A comparison of earthquake acceleration with intensity ratings, Bull. Seism. Soc. America., 46, 317, 1956.

32. Esteva L., Rosenblueth E., Espectros de temblores a distancias moderadas y grandes, Primeras Jornadas Chilenas de Sismol. e Ingen. Antisism., 1, BI.5, 1963.

33. Esteva L., Seismic risk and seismic design input, in R.J.Hansen's Seismic Design of Nuclear Power Plants, M.I.T. Press, 438, 1970.

34. Mickey, W., Strong motion response spectra, Earthquake Notes, 42, 5, 1971.

35. Kanai, K., Improved empirical formula for characteristics of strong earthquake motions, Proc. Japan Earthq. Eng. Symposium, 1, Tokyo 1966.

54

36. Denham D., Small G., Strong motion data centre, Bull. New Zealand Soc. Earthq. Eng., 4, 17, 1971.

37. Cloud W., Perez V., Unusual accelerations recorded at Lima Peru, Bull. Seism. Soc. America, 61, 633, 1971.

38. Krishna, J., Seismic zoning of India, Roorkee Earthq. Seminar, 35, Roorkee, 1959.

39. Oldham, R., The depth of origin of earthquakes, Quart. Journ. Geol. Soc. London, 82, 67, 1926.

40. Blume, J., Earthquake ground motion and engineering procedures from important installations near active faults, Proc. 3rd World Conf. Earthq. Eng., 3, IV-53, 1965.

41. Housner, G., Intensity of earthquake ground shaking near the causative fault, Proc. 3rd World Conf. Earthq. Eng., 1, III-94, 1965.

42. Nuttli, O., The Mississippi Valley earthquakes of 1811 and 1812, Bull. Seism. Soc. America, 63, 227, 1973.

43. Sponheuer, W., Untersuchung zur Seismizität von Deutschland, Veröff. d.Inst. Bodendynam. Erdbeben-forsch., 72, 23, Jena 1962.

44. Neumann, F., A broad formula for estimating earthquake forces on oscillators, Proc. 2nd World Conf. Earthq. Eng., 2, 850, Tokyo 1960.

45. Ambraseys N., Behaviour of foundation materials during strong earthquakes, Proc. 4th European Symp. Earthq. Eng., London, (summary p.4), 1972.

ANALYSIS OF SEISMIC RISK

by M.Caputo
Istituto di Geofisica,
Università Bologna, Italy.

INTRODUCTION

Historical and observational data on seismic events
in the Italian regions have been studied and analyzed
for evaluation of seismic risk and zoning. A new technique is presented whereby the available data are classified and ordered for each region subdivided into small
cells of equal areas. The results are represented by a
map obtained through electronic computation and a plotter. Low pass filtering of the numerical results is
done to gradually give improved seismic zoning of the
regions.

Magnitude-frequency relations of the regions in
conjunction with the above seismic zoning are investigated, and the stability of the parameters of these
relations is studied for each region. Energy release
mechanism connected with the number and areas of faults
in each region is also investigated and used to
strengthen the supposition of seismic stability of the
regions proposed.

The above investigation is then reviewed from the
point of view of seismic energy propagation. In this
context, various forms of energy propagation and release due to earthquakes is described.

Finally, the evaluation of seismic risk in each
region is derived by simple statistical approach and

formulae for the frequency of occurence of maximum magnitude values and their return periods are obtained for
definition of the seismic risk.

In the above described investigations, the set of
information needed for each separate part has been summarized below:

- For the probability of earthquakes, a set of events
  represented in a $S_5(x_i)$:

  x1 = longitude
  x2 = latitude
  x3 = time
  x4 = magnitude
  x5 = maximum intensity

- For the propagation of the seismic energy, a set of
  events represented in a $S_7(x_i)$:

  x_1 = longitude of event
  x_2 = latitude of event
  x_3 = elongation of isoseismals of various orders
  x_4 = semi-major axis of isoseismals
  x_5 = azimuth of semi-major axis
  x_6 = duration of the event
  x_7 = magnitude

- For the estimate of the risk of damage at a point, a
  set of parameters represented in a $S_5(x_i)$:

  x_1 = longitude of the point
  x_2 = latitude of the point
  x_3 = value of the point
  x_4 = vertical component of accelerations
  x_5 = horizontal component of accelerations

This information can be turned over to engineers
to establish the risk of damage in a region for a given
period. Thereby, the engineers can plan, according to
some optimization principles, for the safety of the
value of their investments.

The discussion of this problem should be made in
conjunction with experts in seismic engineering, in
archeology, in museums conservation, sociology, etc.
These aspects are not within the scope of this lecture;
therefore only a brief description of methods to be used
to estimate the risk will be given.

SEISMIC ZONING

The data can be collected from two different
sources. The data collected from seismic bulletins or
directly from the seismic stations, which are recorded
instrumentally, are called instrumental data. All the
other data are called historic. They are collected
from chronicles, records of community houses, history
books, newspapers, etc. Both data, although with different accuracy, contain information on the coordinates
of the epicenter, magnitude or intensity, time and
duration of the event. The set of data is called a
catalogue.

Obviously, to study the seismicity of a given
region, one should have a time series of events as long
as possible.

Since the method of representing the data is most
important, the first analysis of accuracy is that of
the coordinates. This accuracy sets a lower bound on
the sides of the cells of the micro-zoning to be used
for representing the data on a map. In Italy, about
5000 events were collected from the year zero to 1970.
About 90% of these have coordinates which have a standard deviation of 10 kms; therefore a micro-zoning mode
was used with squares of sides of 10 kms, taking into
account the fact that in each of these squares, in most
of the seismic areas,one has at least 5 events.

The next analysis is for the completeness of the
catalogue. Obviously one may expect that for each region events with larger magnitude are more likely to be
present in the catalogue, also that the probability is
a decreasing function of time counted backwards from
the end of the catalogue. For these reasons one should
establish the time interval for which all the events of
a fixed magnitude which actually occurred are recorded,
thus leading to the determination of a function:

$$M = f(t) \qquad , \qquad f'(t) < 0$$

where t is the lower bound of that interval. Practically, for the Italian catalogue the matrix shown in
Table 1 was drawn where the heavy line gives the time
interval in which the catalogue is complete for a fixed
magnitude. The line has been drawn considering for
each magnitude a subdivision of the catalogue into a set
of equal time intervals, of lengths containing a sufficient number of events, (therefore not necessarily the

same length for all magnitudes), and defining the time
where there is a sharp and stable change in the number
of events for each interval. Luckily there has never
been any doubt in establishing this time. In the table
of the matrix the length of the interval has been
standardized for all magnitudes for the sake of simplicity of representation. Repeating this procedure for
each seismic region it can therefore be assumed that
in the periods in which the catalogue was completed,
for each magnitude there was stability af the seismic
activity.

Obviously these criteria could be formulated more
accurately from a statistical and mathematical point of
view, for instance to establish the confidence level of
the statements, to determine more accurately the intervals of completeness of the catalogue, etc. Instead of
spending time in obtaining results which are exact from
mathematical and statistical points of view, it is first
preferred to survey and do a feasibility study of the
steps to be taken to reach the necessary results.

Thus one can proceed to the next step which is the
representation of the catalogue on a map, Figure 1, for
the purpose of seismic zoning. This can be done with
an electronic computer and a plotter. The significant
data are put on a magnetic tape and ordered to obtain
the number of events for each square of the microzoning,
Figure 2. This representation can be filtered with a
low pass two-dimensional filter. After a few trials to
establish the cut-off frequency and number of weights,
which depend on the data, one may proceed to select the
proper filter and to plot the filtered data; the result is a number for each square of the zoning indicating the number of events occurring in that square. The
purpose of the filtering is to eliminate the background
noise and smooth the data on account of the errors in
the coordinates and number of the events.

Figure 3 shows the filtered data, where contours
of the areas which have the same number of events per
square are drawn. The selection of the filter depends
on the quality and quantity of the data, and on the
size of the structure of the zoning to be obtained. Of
course a great number of the weights of the filter help
in keeping the distortion introduced by the filter at a
low level, but it decreases the area of the significant
part of the filtered map. In order to satisfy both requirements a balance should be found. Also the cut-off

	3	4	5	6	7	8	9	10	11	Total
1885	3.0	4.0	1.0	23.5	19.0	4.0	2.5	0.0	0.0	57.0
1890	5.0	3.0	3.0	19.0	9.0	4.0	2.0	1.0	0.0	46.0
1895	0.0	1.0	3.5	43.5	29.5	9.5	3.5	.5	0.0	91.0
1900	10.5	9.5	5.0	45.5	14.5	6.5	1.5	0.0	0.0	93.0
1905	2.0	17.0	11.5	73.0	36.5	8.5	4.5	0.0	1.0	154.0
1910	3.0	11.0	16.0	144.0	48.5	11.5	3.0	1.0	1.0	239.0
1915	0.0	3.0	.5	81.5	20.0	2.0	2.0	2.0	0.0	111.0
1920	27.0	33.0	28.0	56.5	26.5	7.5	3.0	1.5	0.0	183.0
1925	2.5	11.0	16.5	57.5	18.0	3.5	0.0	0.0	0.0	109.0
1930	23.5	53.0	26.5	58.0	15.5	3.5	2.5	1.5	0.0	184.0
1935	8.5	18.0	22.0	33.5	15.0	1.0	1.0	0.0	0.0	100.0
1940	2.0	13.5	18.5	43.0	12.5	6.5	1.0	0.0	0.0	97.0
1945	4.5	11.0	14.5	21.0	11.5	1.5	1.0	0.0	0.0	65.0
1950	2.5	9.0	11.5	34.5	16.5	5.0	1.0	0.0	0.0	80.0
1955	15.0	32.5	50.5	36.5	12.5	2.0	0.0	0.0	0.0	149.0
1960	105.0	118.0	90.5	41.5	18.5	2.5	0.0	0.0	0.0	408.0
1965	59.5	123.5	64.0	46.5	12.5	9.5	1.5	0.0	0.0	351.0
1970	60.0	80.0	47.5	29.0	9.0	5.0	2.5	0.0	0.0	266.0

Table 1: Completeness of the Catalogue

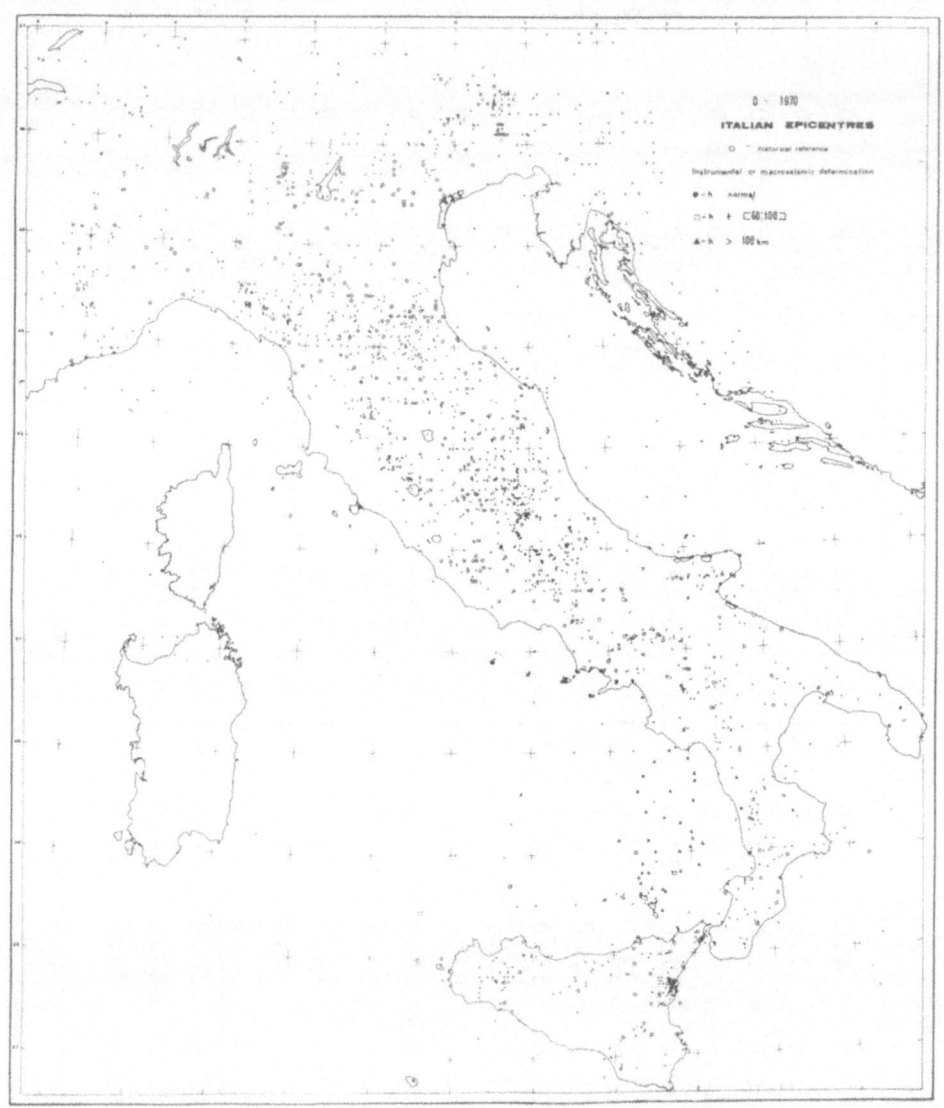

Figure 1: Catalogue of earthquakes in Italy

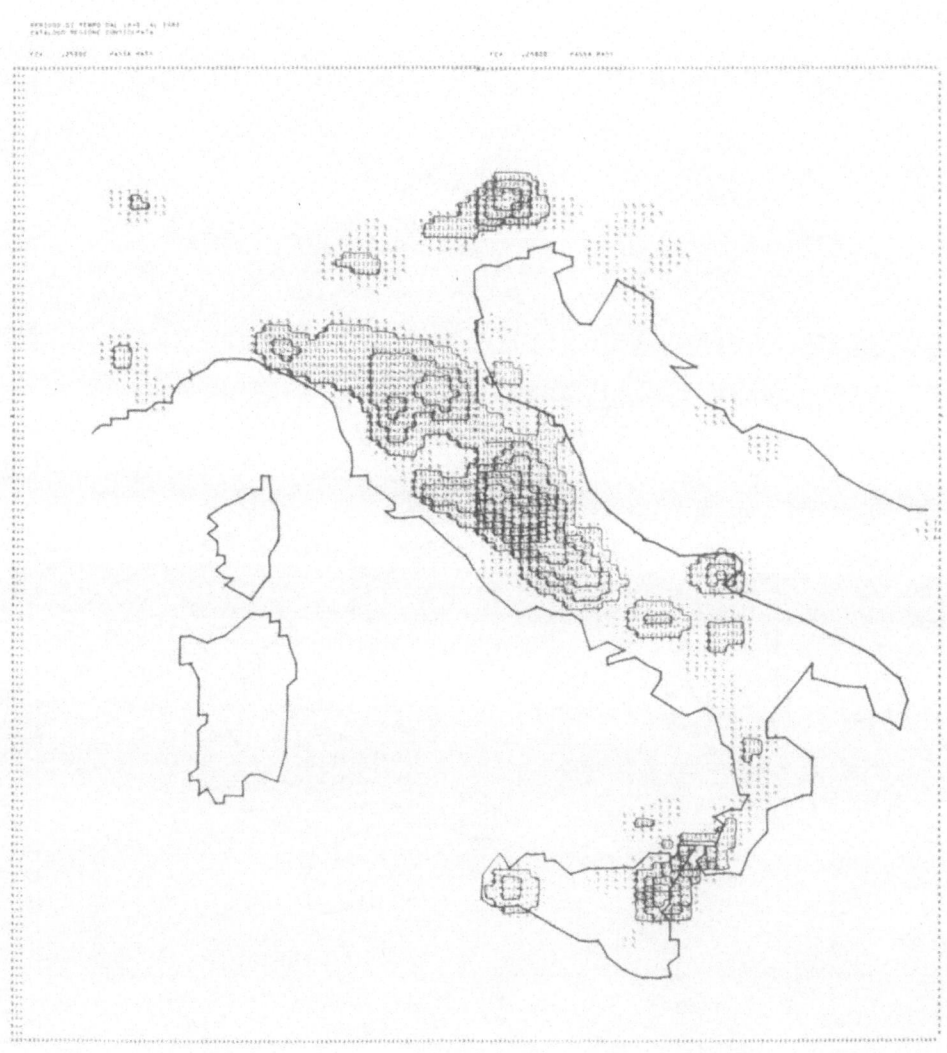

Figure 2: Representation of the data; number of events
per square.

62

Figure 3: Filtered data of Figure 2 and tentative seismic zoning based on seismic information.
$f_c = \frac{1}{4} f_n$.

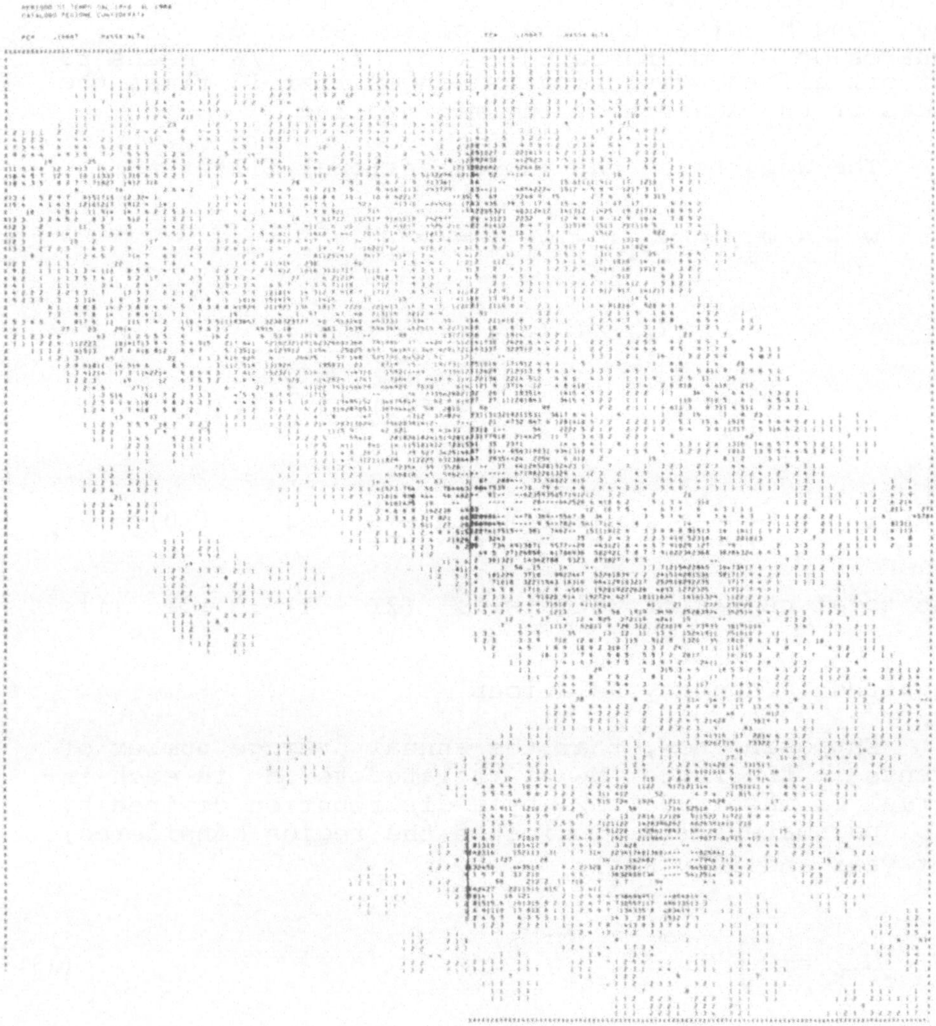

Figure 4: High pass filtered data of Figure 2, $f_c = \frac{1}{4} f_n$.

frequency should be established to leave enough energy in the signal and still eliminate a good part of the noise.

The two-dimensional filter used is a low pass filter with cut-off frequency $f_c = 1/4$ measured in units of the Nyquist frequency. The Nyquist frequency has a wave length twice the length of the sides of the squares of he microzoning; e.g. $f_c = 1/6$ means to cut off all the wave lengths longer than 12 times the sides of the squares of the microzoning.

The weights of the two-dimensional filter are:

$$w_{ij} = w_{ji} = w_{-ij} = w_{i-j} = w_{-i-j}$$

$$w_{ij} = v_i v_j$$

$$v_i = f(i)/\beta \qquad\qquad \beta = f(0) + 2 \sum_1^I {}_n f(n)$$

$$f(i) = \frac{2(I+1)}{(\pi i)^2} \sin(\pi i f_c) \sin\frac{\pi i}{I+1} \quad ; \qquad f(0) = 2f_c$$

The total number of weights is $(2I + 1)^2$; i.e. $|i| \leq I$.

Magnitude-Frequency Relations

Supposing now, that the annual average number of events in the range M_k is related to M in each interval of magnitude M_k by a distribution defined by N_{ik} parameters (i indicates the region considered; per year and 1000 Km^2):

$$N_{ik} = \int_{M_k} 10^{\alpha_i - \gamma_i (M-M_0)} dM \qquad\qquad (1)$$

where γ_i gives the relation between the number of small magnitude earthquakes and those of large magnitude. The larger γ_i is, the more the system is dominated by small earthquakes. α_i (together with $\gamma_i M_0$) may be considered a coefficient of seismic intensity, M_0 may be chosen arbitrarily, but it is customary to assume $M_o = 8$.

From the α_i and γ_i parameters one can go back to the return periods of the earthquakes that have a magnitude greater than a pre-established magnitude M , or the maximum magnitude observed annually with the maximum frequency, as will be seen later.

In fact from (1), integrating between M and $M + \Delta M$,

$$\Delta N_{iK} = \frac{e^{[\alpha_i - \gamma_i(M-M_o)]\ln 10}(1 - e^{-\gamma_i \Delta M \ln 10})}{-\gamma_i \ln 10}$$

$$\left. \begin{array}{l} = e^{[\alpha_i - \gamma_i(M-M_o)]\ln 10} \Delta M \\[2ex] \ln \Delta N_i = \alpha_i \ln 10 - \gamma_i(M-M_o)\ln 10 + \ln \Delta M \end{array} \right\} \quad (2)$$

$$\log \Delta N_i = \alpha_i - \gamma_i(M-M_o) + \log \Delta M$$

from which follows the well-known and accepted law:

$$\left. \begin{array}{l} \log \Delta N = a + b\,M \\[2ex] a = \alpha_i + \gamma_i M_o + \dfrac{\ln \Delta M}{\ln 10} = \alpha_i + \gamma_i M_o - \log \Delta M \\[2ex] b = -\gamma_i \end{array} \right\} \quad (3)$$

At this stage, one could proceed to estimate the parameters α_i and γ_i for instance with the least squares method considering errors in M and N . In each seismic area 10^a should vary linearly across the various contours and b should stay constant. The change of b from one seismic area to another should imply different mechanisms of energy release and different capability of energy accumulation. This analysis gives information on the energy release in the zones of the seismic regions, which have the same number of events per unit area and unit time.

The seismic risk can be computed in the same man-

ner. One should use only the events of a fixed magnitude M , computing N(P,M) for each contour and unit
area, then $a_i(P)$ and $b_i(P)$ by least square interpol-
ation can be estimated.

Stability of the α and γ parameters

The analysis presented in the preceding Section
can also be made in a different way which will be presented in this Section. First the problem will be re-
formulated.

Two catalogues of earthquakes were considered; one
with about 985 "instrumental" events from 1893 to 1965
of 4.2 to 7.6 magnitude, and one with about 252 "historical" events from 1501 to 1929 of IX and X intensity.

Italy was divided into three regions 1, 2, 3 as is
shown in Figure 5. This subdivision was made in an attempt to obtain statistically uniform regions. The results of this subdivision should be interesting from
the point of view of earthquake risk understood in an
overall sense for a sufficiently large number of objects
distributed uniformly throughout the region.

The analysis of the data was effected by various
methods. In Table 2, the method that concerns this
research will be presented; the quantities calculated
are given at the head of each column of the table itself.

The calculations were made eliminating the aftershocks in the instrumental catalogue. An earthquake is
considered an aftershock in the instrumental catalogue
when it occurs within a week of the previous one and
the epicentral distance between the two is 10 Km. In
the historical catalogue the time interval limit was
increased to two weeks, and the distance was replaced
by the region.

Variations in α_i mean that the intensity of seismic activity varies from region to region.

In order to analyse the stability of seismic activity in time, one proceeds as follows. From the α_i
and γ_i parameters of the instrumental catalogue the
numbers of instrumental earthquakes of IX and X intensity in the three regions are calculated. These numbers

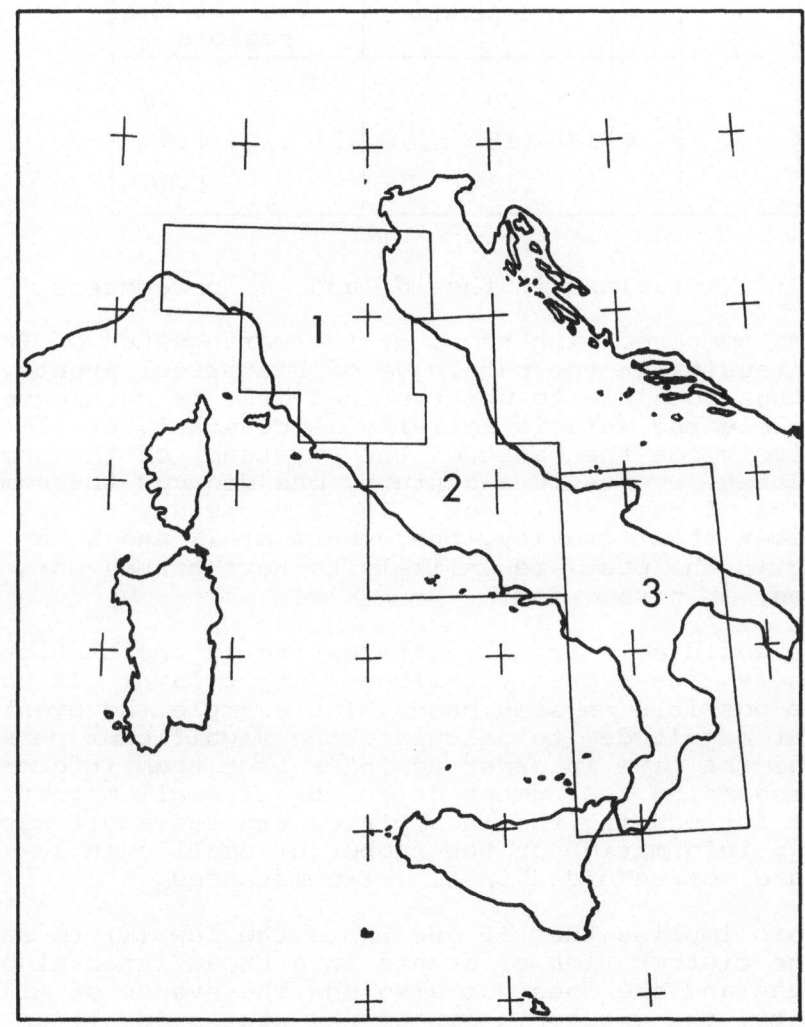

Figure 5: Tentative choice of seismic regions in Italy as a working hypothesis.

Region	M_K	γ	$\Delta\gamma$	$\gamma M_o+\alpha$	γ	$\Delta\gamma$	$\gamma M_o+\alpha$
		α and γ are independent			γ is the same for all the regions		
1		.75	.20	1.75			1.70
2	.4	.74	.15	1.94	.74	.10	1.92
3		.74	.13	1.86			1.86

Table 2: Variations in the α and γ parameters

are then compared with those of the earthquakes of IX
to X intensity in the catalogue of historical events.
It is thus possible to observe that if this catalogue
is complete the seismic activity is constant, or alternatively, on the basis of the constancy of the number of these events in a century, one deduces the completeness of the catalogues. It is necessary to say
that given their gravity, the events of IX and X intensity in the past are unlikely to have passed un-
observed, or unrecorded.

It would be extremely interesting if the statistics
of these events strictly followed simple laws. It would
then be possible in some cases, for example the events
of great magnitude, to calculate the significant parameters of the laws in order to infer from them information regarding the number of events of small magnitude,
without introducing the incomplete, and therefore misleading, information on the number of small events
which are not recorded in all circumstances.

This implies that if one wants the law (1) to satisfy the distribution of events in a broad interval of
magnitude and one does not also use the events of minor
intensity, the extrapolation of the statistics leads to
an underestimate of the number of events of great magnitude.

Therefore the catalogues with historical events,
though extremely useful because they also serve to
demonstrate seismic stability and complete the statistics for the events of IX and X intensity, do not yet

serve the purpose of improving the properties of extrapolation towards the small magnitudes.

From this conclusion there follows another for the
faults, in fact the accumulation of energy W which is
in part transformed into seismic energy E is characteristic of the single regions; so the energy of the
earthquake is linked to the dimensions of the faults by
the relations:

$$W = (H/L)^{\frac{1}{2}}(Kp^2 S^{3|2}/\mu$$
$$W = (Kp^2 S^{3|2})/\mu \quad , \quad E = \eta W$$
$$\left.\right\} \tag{4}$$

Where K is a parameter near to 1 , which depends on
the model, η is a parameter comprised in the interval
$[1/3, 1]$; μ is the modulus of rigidity.

Of the expressions in (4) the first is true for
rectangular faults of L   length, H   depth and S area,
and the second is true for circular or square faults of
S area. It follows that the ratio between the number
of large faults and that of the smaller ones should be
fairly constant in the various regions on the hypothesis
that the difference p between the state of stress before and after the earthquake is constant in Italy
given that μ , the modulus of rigidity, may be assumed
to vary little.

Linear elasticity of the anistropic means govern
the properties of the isoseismals. Moreover, it has
been proved that the following empirical relation is
true in a vast interval of M

$$\log E = A + BM \tag{5}$$

with $A = 11.8, B = 1.5$ which associated with the expressions (4) provides the following formulae:

$$\log S = (1/3)\log(L/H)+(2/3)(A+BM)-(2/3)\log(\eta Kp^2/\mu)$$
$$\tag{6}$$
$$\log S = (2/3)(A+BM)-(2/3)\log(\eta Kp^2/\mu)$$

Considering certain limits for the typical dimensions of the faults in Italy, it was calculated that the
preceding formulae are true for $4.5<M<7.5$.

For the three Italian regions indicated in Figure 5, the values corresponding to α_i and γ_i are given in Table 2. This regionalization has greater interest as regards earthquake risks. One can see confirmed in the three regions the stability of the ratio between the numbers of large and small earthquakes in support of what has been said above in connexion with the stability of the distribution of the faults according to their energy and geometric characteristics.

That is, if the faults are active in the same manner, then the number of faultings and faults in the Italian regions according to their S areas, less a constant which depends on the average number of events per fault, is still given by the expression in (1).

Concluding this first investigation, it can be said as an initial approximation, that there are not large differences of a seismic character among these regions apparently so different geologically, at least as regards the features that regionalization intended to distinguish. These regions are also found to be seismically stable in the period from 1500 to today.

In Table 2, it may be noticed that the values of γ are almost the same in the three regions 1, 2, 3 of Figure 5. This could imply that the geometry of the mechanism of seismic energy release in the three regions is the same. To check this result, another subdivision of Italy into four regions may be considered as shown in Figure 6. Table 3 gives the values of γ

	γ	$\Delta\gamma$	$\gamma M_o+\alpha$	γ	$\Delta\gamma$	$\gamma M_o+\alpha$
	α and γ are independent			γ is the same for all regions		
CD	.67	.24	1.26			1.96
CL	.71	.19	1.23	.77	.30	1.51
S	.89	.32	2.03			1.44
N	.98	.39	2.48			1.49

Table 3: The α and γ values for a four region subdivision of Italy.

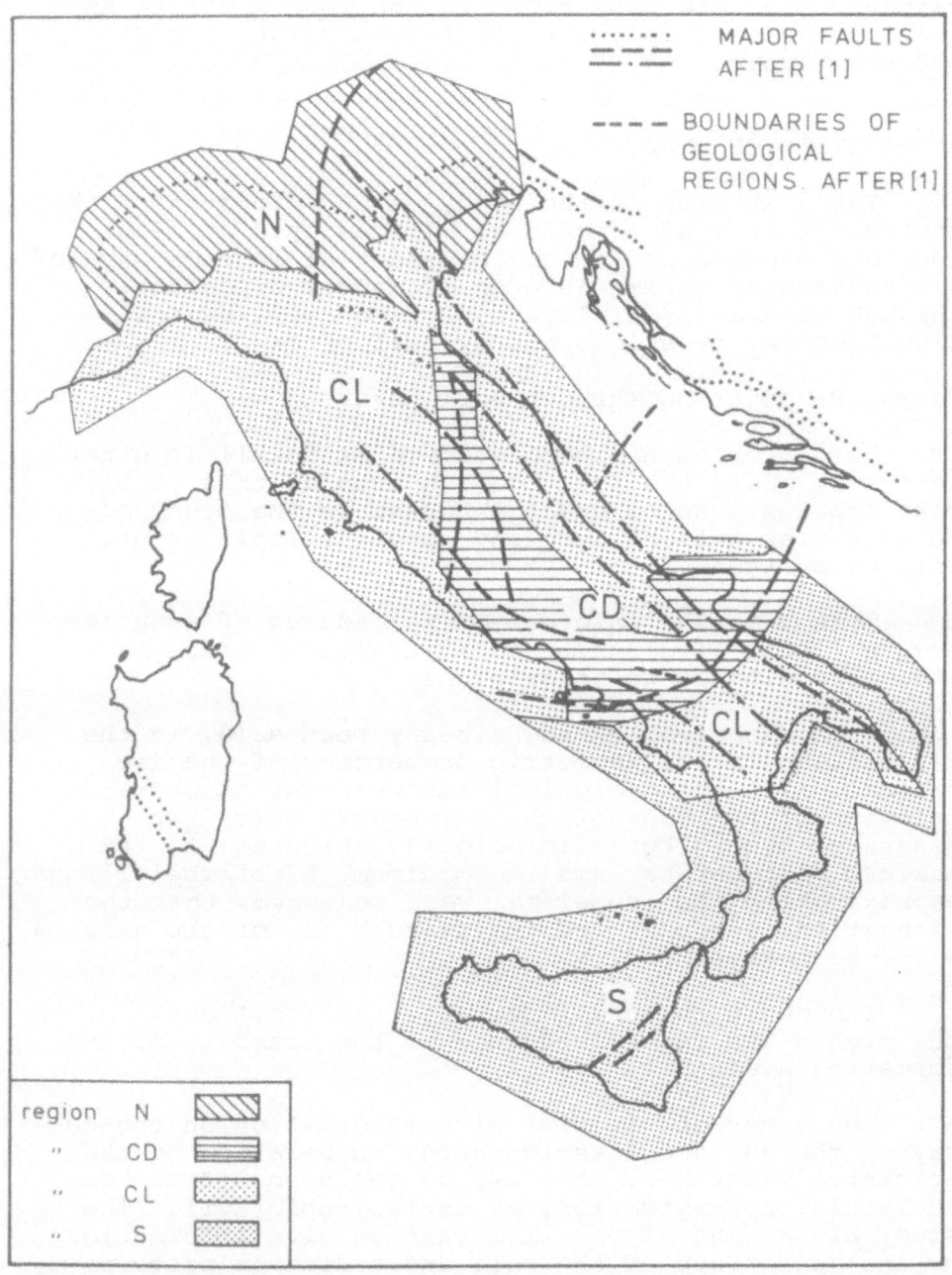

Figure 6: Tentative choice of seismic regions in Italy according to geological information on fault-

and α for the four regions and one may see that the
values of γ are very close in the region of the Ap-
pennines while they assume different values for Sicily
and the Alps.

PROPAGATION OF ENERGY

The next step is the study of the propagation of
energy. This study should be divided into at least 5
sections because one should consider at least 5 ways of
propagation of seismic energy or causes of damage:
through or along soil layers, by tsunamis, by soil
liquefaction, by slides and by fall of dams.

Energy Propagation Through Soil Layers

This problem has been studied in detail in other
lectures. Here the practical aspects of this problem
with implications of great interest in the study of
seismic risk will be seen and single seismic events
will be analysed.

This time the information is made up of about 69
events (Table 4) distributed in Italy, of known magnitude and with known isoseismals.

The objective, as has already been said, is the
correlation of the geometric properties of the isoseismals with the geological and geophysical characteristics of the zone of the earthquake which has
generated them. For this purpose, the area $\bar{A}$, the
maximum diameter a and the minimum b of the ellipse
approximating the isoseismal were measured, then the
elongation e = b/a and the azimuth α of the axis a
with the nearest fault.

Though it is very important, the focal depth of
the events was ignored because of the scarcity of information available.

The study of $\bar{A}$ will give information on the decay of the surface elastic energy in relation to the
epicentral distance, that is, on the supposition, as
an initial approximation, of an isotropic soil. The
study of e and of α will instead give information
on the anisotropy of the soil and radiation pattern
of the event with qualitative and quantitative charact-
ers.

Date	ℓ				Epicentre	
	7	8	9	10	λ	φ
2-1703			.48	.45	13.5	42.3
3-1783			.51	.63	15.1	38.3
3-1783			.50	.44	16.5	38.9
1-1832	.95	.56			12.7	10.7
5-1898	.68	.75			12.8	47.4
7-1901	.88	.64			18.7	41.6
2-1904	.81	.76			18.3	42.0
10-1907			.55	.67	16.1	38.0
8-1909	.58	.49			11.2	43.1
5-1910	.52	.23	.21	.17	15.4	40.9
11-1916	.73	.71			13.2	42.6
4-1917	.61	.34	.50		12.1	43.5
9-1915	.53	.41	.48		11.5	49.8
9-1919	.78				11.7	42.8
9-1920	.56	.56	.48		10.3	44.2
12-1927	.75				12.7	41.7
7-1930	.58	.54	.45	.30	15.4	41.0
12-1936	.82	.78			18.2	41.1
7-1937	.83	.83			15.4	41.7
10-1940	.52	.57			11.7	42.9
10-1943	.46	.53			13.6	43.0
6-1948	.85				12.2	43.5
10-1961	.48	.43			15.9	40.2

Table 4: Epicenter location and isoseismal characteristics of 69 earthquakes in Italy.

As has been seen, the energy W of the earthquake
is partly transformed into elastic energy E ; the
relationships (6) and (5) are true from which it follows

$$\log E \sim \log S \qquad\qquad\Big\}$$
$$\log E = A + BM, \ \log S = N + kM \Big\} \qquad\qquad (7)$$

of which the first derives from (6) even if in an indicative manner as will be seen, the second is exper-
imentally true for a vast interval of M and the third
expression is the consequence of the preceding expressions. Or alternatively, indicating by N a parameter
which is easily calculated, one has, for circular and
square faults or for rectangular faults respectively

$$\log S = N + kM \qquad\qquad\Big\}$$
$$\log S = (1/3)\log(L/H) = N + kM \Big\} \qquad (8)$$

If for a given value of M for the Italian region only
L increases, then also for the rectangular faults one
may write for M greater than a certain value

$$\log S = P + QM \qquad\qquad\qquad (9)$$

In order to write the expressions (8) it was supposed
that $\eta Kp^2/\mu$ did not undergo large variations in the
regions considered. These formulae may serve to carry
out a test on the variability of $\eta Kp^2/\mu$ and on the
shape and dimensions of the faults.

Reasonably assuming $K = 1$, $\mu = 5.10^{11}$ and $\eta = \frac{1}{2}$
the first of the expressions (4), from (6) may be
written

$$S = p^{-4/3}10^{15.6+M} \qquad\qquad\Big\}$$
$$M = -15.6 + \log S + (4/3)\log p \ . \Big\} \qquad (10)$$

The value of $p = 26.10^6$ was estimated for events
in the S.Andrea region, for two other earthquakes in
Colorado the estimate was instead 3.10^6 and 22.10^6 .
Assuming tentatively the values $p = 3.10^6$ and
$p = 30.10^6$,

$$M = \log S - 7 \qquad\Big\}$$
$$M = \log S - 5.6 \Big\} \qquad\qquad (11)$$

For the Italian region the formulae (11) might give a rough estimate of the area of the square or circular fault, from the magnitude that is to be expected from a fault of known dimensions.

A working hypothesis based on the law log S=N+kM is that also the area $\overline{A}$ of the isoseismals is proportional to the diameter of the faults and therefore its logarithm is a linear function of the magnitude and thus also its elongation e , that is:

$$\left.\begin{aligned} \log \overline{A} &= C_A(I) + D_A(I)M \\ \log e &= C_e(I) + D_e(I)M \end{aligned}\right\} \tag{12}$$

From the tables of the report mentioned, the values $C_A(I)$ and $D_A(I)$ were taken for the values of I given in the columns of Table 5.

INTENSITY	LEAST SQUARE METHOD				MAXIMUM LIKELIHOOD			
	Same D_A for all I		Independent parameters		Same D_A for all I		Independent parameters	
	C_A	D_A	C_A	D_A	C_A	D_A	C_A	D_A
VII	-0.738	0.716	-0.60	0.69	-1.24	0.817	-1.31	0.83
VIII	-1.133	0.716	-1.32	0.75	-1.67	0.817	-1.80	0.84
IX	-1.616	0.716	-1.64	0.72	-2.32	0.817	-1.96	0.77
X	-2.170	0.716	-2.13	0.71	-2.74	0.817	-2.16	0.72

Figure 5: Variation of C_A and D_A with the intensity I .

It can be seen that D_A does not vary significantly and that with a good approximation one can write

$$C_A = \ell + mI \tag{13}$$

Then the values of C_A were calculated for an average value of D_A common to all the values of I . These values are given in the first two columns of Table 5.

Ignoring the flattening of the isoseismals, one

has as an initial approximation

$$\log \pi R^2 = \ell + mI + D_A M . \tag{14}$$

where $\pi R^2 = \overline{A}$ and D_A is the average value of $D_A(I)$ calculated on the assumption that is the same for all values of I .

With the maximum likelihood method one obtains:

$$I = [-\log\pi R^2 + 2.39 + 0.92 \, M]1.94 \tag{15}$$

and with the least square method

$$I = [-\log\pi R^2 + 2.65 + 0.72 \, M]2.08 \tag{16}$$

which give a fairly good approximation, apart from the anisotropy, of the decay of intensity I with the distance R for $4.2 \leq M \leq 7.6$ and $7 \leq I \leq 10$.

This result is of great interest in connexion with investigations into the dissipation of elastic energy near the source.

The results also confirm indirectly that the fall in stress before and after the earthquake probably does not vary much in the various regions and for the various magnitudes. Another interesting relation is obtained rewriting the second of (4):

$$M = (1/B)(-A + \log(S^{3/2}\eta p^2 K/\mu)) \tag{17}$$

which, associated with (14), provides

$$I = (1/m)[\log\pi R^2 - e - (D_A/B)\{-A + \log(S^{3/2}\eta p^2 K/\mu)\}] \tag{18}$$

which links the area of a fault to the intensity and the epicentral distance. Passing to the numerical values of (11) and (15) the values of ℓ, m, D_A of the maximum likelihood method, one obtains for S in Km^2

$$I = -1.89[\log\pi R^2 - 2.39 + 0.82(-5.6 + (4/3)\log p - \log S)] \tag{19}$$

which provides a rough estimate of the intensity, as a function of the distance, associated to a fault of area S that it is supposed to cause earthquakes.

It is clear that for satisfactory use of (10) and (19), the parameter which needs a better estimate is p; in fact an error of a factor of 10 in the estimate of p causes an error of 2 units of intensity or an error of 1 in the magnitude.

It is known that isotropic bodies cease to be such under stresses with non-diagonal tensor. The stresses that cause deformations of the crust create a certain anisotropy of the medium beyond the immediate vicinity of the faults, which already constitute a considerable local anisotropy. This investigation with the study of e is intended to serve as an introduction to an investigation of the anisotropy of the crust.

In fact $\overline{A}$ provides information of mean isotropic nature, e can provide, as will be seen, information on the anisotropy of the medium or the radiation pattern with a quantitative character, α might supply further information of a quantitative nature on such anisotropy or radiation pattern.

It has been said that the crust, because of tectonic stresses, might be in a state of anisotropy. In this case, if the radiation pattern is spherically symmetric, the flattening of the isoseismals must be independent of e . The following considerations constitute a test of this state of stress.

Consider the following relations:

$$\log e = C_e(I) + D_e(I)M$$

and the calculated values of $C_e(I)$ and $D_e(I)$ given in Table 6.

I	$C_e(I)$	$D_e(I)$
VII......	0.098	−0.065
VIII.....	0.381	−0.115
IX.......	0.222	−0.089

Table 6: Variation of C_e and D_e with intensity I .

It can be seen that $D_e(I)$ is always negative; this means that the flattening, which always varies between 0 and 1 , increases with magnitude. This is reasonable since, as has been seen, the magnitude is linked to the dimensions of the faults, and it is also reasonable that at a certain point the size of the faults should increase preferably in length and so cause a greater flattening of the isoseismals. The negative values found for $D_e(I)$ therefore also confirm

the lengthening of the faults with an increase in S.
It can also be seen that $C_e(I)$ and $D_e(I)$ do not
appreciably vary with variation in I^e and that the
values $D_e(I)$ are very small. Bearing in mind the low
accuracy of the data and above all the smallness of
their number, one might also think that the variations
in $C_e(I)$ and $D_e(I)$ form part of the statistical
fluctuations. These results might suggest that the
flattening of the isoseismals is in part due to an anisotropy of the crust in the regions examined.

An examination of the flattening of the isoseiamals
associated with each single earthquake given in Table 6,
for the time being confirms that there is no evidence
that the flattening varies monotonically with the intensity of the isoseismals and moreover the scarcity of
the data does not suggest any quantitative analysis.

If there is anisotropy in the crust, the direction
of the major axis of the isoseismals might serve to indicate the directions of minimum and maximum dissipation
of the elastic energy and therefore the principal axes
of the stress causing the anisotropy according to the
theory of linear elasticity. Calculations should make
it possible to give information also on the extent of
the stress to which the crust is subjected.

The examination of the correlation between the
direction of the axis a of the isoseismals and that
of the nearest local geological structures seems to
confirm the hypothesis of anisotropy of the crust, in
the sense mentioned above.

Tsunami

An important effect of earthquakes is that of
tsunamis. They are associated with earthquakes in the
sea, which cause the motion of the sea bottom; this in
turn can cause large sea waves which may reach the coast
and break on it reaching elevations of 30 metres or
more with devastating effect on all construction and
life in inhabited areas.

The problem needs a separate discussion; here the
attention is only called to the importance of this
phenomenon and to its risk which involves many more
coastal lines than one may think.

As an example, the 1960 Chilean earthquake gener-

ated a tsunami which reached the coast of Japan 13,000
Km away. There is sometimes a long enough time interval
between the earthquake and the breaking of the wave on
the coast to allow a warning to be given, thus saving
many lives.

Soil liquefaction

Another phenomenon which is of great importance in
many areas is that of soil liquefaction. During an
earthquake, because of the vibrations, non-compacted
saturated sand looses its water content which flows
towards the surface of the soil and causes the formation of small dunes. During this phenomenon the soil
turns into quicksand and constructions or vehicles on
it may sink or float according to their specific weight.

Sometimes this phenomenon occurs on one side of a
building only. Masonry type buildings may collapse
completely, and reinforced concrete structures may be
toppled and found lying on the ground without much
damage as for instance occurred in Niigata, Japan 1964.
The tenants of the building left it through the windows
and walking on the walls. Other buildings sank about
three metres.

Soil liquefaction often causes landslides when it
occurs in a saturated sand layer under a layer of soil
on a slope. These slides are called flow slides; the
largest of them is that of 1920 in the Province of Kan-
su. It covered an area of 200 square Km., translated
houses for a about 1 Km. and buried whole cities.
100,000 lives were lost.

This problem, as well as the following one, needs
a separate treatment; in this context they can only be
mentioned.

Soil liquefaction under dams and roads or railways
may cause their sinking or translation. It is well
known that a 100 m long section of the Sheffield Dam in
California during the 1925 earthquake slid 100 m down-
hill. During the 1960 earthquakes, a Chilean road
built across a swamp disappeared into the water, while
before it was 1.2 m above it.

Slides

Clay can also slide during earthquakes although

there is no evidence that clay can be subject to liquefaction.

(One of the largest slides is that which occurred
at Anchorage in Alaska during the 1964 earthquake; it
covered about 5 square Kms., some houses were translated 200 m, and the ground was lowered by an average
of 10 m.)

THE SEISMIC RISK

To estimate the seismic risk in a given seismic
zone, the statistics will now be formulated in a simpler
form. Consider the series of events in time with magnitude $M > M_O$, and let α be the average number of
such events per unit time; then the time series of
events can be represented by a Poisson distribution

$$f(n, \Delta t) = \frac{(\alpha \Delta t)^n}{n!} e^{-\alpha \Delta t} \tag{20}$$

$f(n, \Delta t)$ is the frequency of a set of n events in the
interval Δt (or the probability of having n events
of magnitude $M > M_O$ in Δt). The distribution of the
magnitude is given by (1) which shall be rewritten as
follows

$$N_{ki} = \int_M \alpha e^{-\gamma M} \, dM \tag{21}$$

N_{ki} is the annual average number of the events of
magnitude in the range M_k per unit area. For the
number of events of magnitude M in the interval
$(0, x]$ is obtained:

$$N = \int_0^x \gamma \alpha e^{-\gamma M} dM = \alpha \left(1 - e^{-\gamma x}\right) \tag{22}$$

where α and γ are new parametric values different
from those used previously.

Before proceeding, the following hypotheses may be
recapitulated: a) The number of earthquakes per annum
is a random variable of Poisson type with average α (see
(20) with $\Delta t = 1$ year)., b) The magnitude is a random

variable given by (21).

The cumulative distribution (c.d.f.) of the magnitude is therefore (remember that $\lim\limits_{x\to\infty} N = \alpha$ is the total number of events)

$$F(x) = P[0<M\leq x] = 1-e^{-\gamma x} \tag{23}$$

and therefore the average annual number of events with magnitude $M > x$ is

$$N_x = \alpha P[M>x] = \alpha e^{-\gamma x} \tag{24}$$

To compute α and γ proceed as follows.
As a consequence of (20) with $\Delta t = 1$ it can be seen that the probability of having r events of magnitude larger than M_0 (here it is assumed that $M_0 = 0$) in a year is:

$$\frac{e^{-\alpha}\alpha^r}{r!} \tag{25}$$

The probability that all these r events have magnitude $M \leq x$ is

$$\frac{e^{-\alpha}\alpha^r}{r!}[F(x)]^r \tag{26}$$

and consequently the c.d.f. of the maximum annual magnitude is

$$G(y) = P[0\leq M\leq y] = \sum_{r=0}^{\infty} \frac{e^{-\alpha}\alpha^r}{r!}[F(y)]^r \tag{27}$$

$$= e^{-\alpha}\sum_{r=0}^{\infty}\frac{[\alpha F(y)]^r}{r!} = \exp[-\alpha e^{-\gamma y}]$$

or

$$\ell_n[-\ell nG(y)] = \ell n\alpha - \gamma y . \tag{28}$$

One may easily verify that $dG/dy > 0$, and $\lim\limits_{\gamma\to\infty} G=1$; the formula is not valid for $y = 0$, in fact $G(0)=0$. Comparing with N_y (24) from which it follows that

$$\ln N_y = \ln \alpha - \gamma y \quad , \tag{29}$$

where N_y or $\ln G(y)$ is the average annual number of events with magnitude $M > y$; therefore from $\ln[-\ln G(y)]$ with (28) one obtains $\ln\alpha - \gamma y$, for (29), or $\ln Ny$ which is what was wanted.

To obtain the numerical values of $G(y)$ consider the time series $y_1, y_2 \ldots y_i \ldots y_n$ of the annual maximum magnitudes and order it such that $y_i \leq y_{i+1}$; then the frequency of $G(y)$ of the events of magnitude y_i is:

$$G(y) = \frac{i}{n+1} \tag{30}$$

and α and γ can be computed from (28) by the least squares method. Finally from (29) N_y is then obtained.

The return period of the events with magnitude $M \geq y$ is

$$T_y = \frac{1}{N_y} = \frac{e^{\gamma y}}{\alpha} = \exp[\gamma y - \ln\alpha] \tag{31}$$

Another relevant parameter is the maximum yearly magnitude, which is most frequently observed. This is obtained from such a value $\bar{y}$ that $g(y) = G'(y)$ as a maximum. One finds that

$$\bar{y} = \frac{\ln\alpha}{\gamma} \quad ; \quad N_{\bar{y}} = \alpha \ e^{\frac{\gamma}{\gamma}\ln\alpha} = 1 \tag{32}$$

As $N_{\bar{y}} = 1$, an earthquake of magnitude $\geq y$ can be expected every year. Or $G(\bar{y}) = e^{-1}$, which means that $\bar{y}$ is the maximum yearly magnitude which is exceeded in 63% of all years.

The maximum yearly magnitude which is most frequently observed in T years is, from the first of (33):

$$y_T = \frac{\ln\alpha T}{\gamma} = \bar{y} + \frac{\ln T}{\gamma} \tag{33}$$

One can also find the yearly maximum magnitude y_p which is exceeded by probability p . From (27) this is:

$$\exp[-\alpha e^{-\gamma y_p}] = 1 - p \tag{34}$$

or

$$y_p = \bar{y} - \frac{\ln[-\ln(1-p)]}{\gamma} \tag{35}$$

More generally, the yearly maximum magnitude $y_p(D)$ which exceeded by probability p in D years, is

$$y_p(D) = y_p + \frac{\ln D}{\gamma} \tag{36}$$

Finally, the risk of earthquakes $R_D(y)$, i.e. the probability of an event of magnitude y or more in D years, from (27) is:

$$R_D(y) = 1 - \exp[-\alpha D e^{-\gamma y}] \tag{37}$$

On this basis one may estimate the economic risk; in fact the damage to a given object is a known function of the intensity of the shaking which in turn should be a known function of the distance and magnitude of the event whose probability of occurrance is given above.

The damage to buildings as a function of the intensity of the shaking is a separate problem of an engineering nature which is not within the scope of this lecture. The problem of shaking as a function of the distance and magnitude has been tentatively treated for the Italian region in the Section on energy propagation through soil layers.

Before closing, attention should be given to the risk of using the formulae obtained for magnitudes which are larger than those observed, or more generally outside the experimental range. It is possible that they follow other probabilistic laws. In other words, one may also say that it is not safe to consider events which have periods of return larger than the time interval in which the catalogue is completed.

For the risk of damage to the objects, it should be added that one should consider that the events occur under the object; for the damage to all the objects in the area one could then take into account the decrease of intensity with distance from the epicentre.

Finally one could never recommend enough that the

theories and formulae be used with great prudence; some
areas in fact have quiescent periods, others may change
their activity abruptly.

BIBLIOGRAPHY

1. Caputo, M., Chiarini, A., Pieri, L., Statistical
    evaluation of earthquakes in three sites proposed
    for a nuclear plant. Comit. Nazionale Energia
    Nucleare, RT/ING(69) 14, Roma 1969.

2. Caputo, M., Seismicity and soil response of the
    Italian region, Moderne vedute sulla geologia deg-
    li Appennini, Atti Acc. Naz. Lincei, Roma 1972.

3. Caputo, M., Keilis-Borok, V.I., Kronrod, T.L.,
    Molchan, G.M., Panza, G.F., Piva, A., Padgaetskaja,
    V.M., Postpischl , D., Models of seismicity and
    seismic risk in Italy (in press).

4. Caputo, M., Keilis-Borok, V.I., Kronrod, T.L.,
    Molchan, G.M., Panza, G.F., Piva, A., Padgaetskaja,
    V.M., Postpischl. D., Models of earthquakes occur-
    ence and isoseists in Italy, Annali di Geofisica,
    2, 1973.

5. Caputo, M., Postpischl, D., Seismicity and tenta-
    tive map of seismic risk in Italy, Proc. V.World-
    conf. Earthq. Eng., Rome 1973.

6. Caputo, M., Postpischl, D.,Numerical seismic zon-
    ing and seismic stability, General Assembly of the
    International Association of Seismology and
    Physics of the Earth Interior, Lima, 1973 (in
    press).

7. Caputo, M., Postpischl, D., Preliminary epicentre
    map of Italy, Akademie der Wissenschaften der DDR,
    Zentralinstitut für Physik der Erde, 18, 31, 1972.

8. Caputo, M., The vibration of a plate with a dis-
    sipative memory, Journal of the Acoustical Society
    of America, March 1974.

9. Barbarella, M., Caputo, M., Romagnoli, E., Soil
    response in two Italian regions, in press, 1973.

10. Bozzi-Zadro, M., Caputo, M., Filtri multidimens-
    ionali e loro applicazioni geofisiche, Annali di
    Geofisica, XXI, 287, 1968.

11. Caputo, M., Postpischl, D., Seismicity of Italian
     Region, In press in Structural Map of Italy,
     C.N.R., Roma, 1973.

EARTHQUAKE GROUND MOTIONS: MEASUREMENT AND CHARACTERISTICS

by S.Cherry

Professor of Civil Engineering,
University of British Columbia,
Vancouver, Canada.

SYNOPSIS

The application of earthquake ground motions in
seismic design is noted and methods of securing these
measurements are outlined. The general characteristics
of strong motion records (accelerograms) are examined
and techniques for analyzing earthquake records are explained. The influence of source mechanism, propagation path geology and local site conditions on the
basic ground motion parameters and on recorded motions
is illustrated. Methods of selecting design earthquakes are surveyed and some comments are offered on
the feasibility of using microtremors as a method of
defining dynamic site characteristics. The material
presented is based on information taken from various
sources in the literature; it is intended as a summary
of current knowledge of earthquake ground motions.

INTRODUCTION

The basic engineering approach involved in securing an earthquake resistant structure can be subdivided
into three main categories;
   (1) analysis for seismic response (loads and de-
       formations)
   (2) design to economically and safely accomodate
       the response
   (3) construction and inspection to ensure the de-
       sign objectives are realized.

The present discussion will be concerned with only part
of the analysis procedure - an understanding of the
character of the earthquake ground motions (and the
factors influencing these motions) to which a system is
expected to be subjected during its lifetime.

Earthquake ground motions exhibit the properties
of a random process and, as such, generate complicated
transient vibrations in a structure. This dynamic response is governed by the characteristics of the
structure and the characteristics of the ground motion
which describe the basic input or excitation function
acting on the structure. It therefore directly follows
that the most important first step in any seismic analysis is the determination of the probable and possible
earthquake ground motions to be expected at the site at
which construction is to take place. These motions are
essentially a function of the regional seismicity, the
nature of the source mechanism, the travel path geology
and local site conditions.

It is not possible to predict precisely the actual
earthquake ground motions to which a proposed structure
will be subjected at some future date. As a result,the
designer is compelled to estimate the seismic exposure
and then select the appropriate ground motion for a
particular site. This is normally done in one of the
following ways:
  (1) by direct extrapolation from comparable re-
      cords in the existing catalogue of recorded
      motions
  (2) by using artificially generated earthquake mo-
      tions which have been modelled, in a probabi-
      listic sense, to simulate the nature of real
      earthquake motions
  (3) by modifying existing records to reflect the
      probable characteristics of rock motions.
These different approaches are briefly noted in a subsequent section of this paper.

MEASUREMENT OF STRONG GROUND MOTIONS

Strong Motion Accelerographs

The basic ground motion input data for earthquake
engineering design calculations are the ground acceleration - time history records (accelerograms). The dynamic forces induced in a structure are directly associated with this parameter of the ground shaking; it'

can be shown that the effective seismic load or lateral
inertia force acting on the mass of a system is equal
to the product of the mass and the ground acceleration.
Once this acceleration record is secured the corresponding velocity and displacement functions can be calculated with satisfactory accuracy by direct integration of the measured motion.

The development of instrumentation for securing
earthquake acceleration records dates back to 1932,
when the U.S. Coast and Geodetic Survey's Strong Motion
Seismograph (Accelerograph) was first introduced. The
first accelerograms were obtained in the 1930's and the
information gained from them resulted in a major break
through in earthquake engineering and provided the im-
petus for new and exciting developments in this field,
developments which are continuing at an accelerated
rate. Contemporary accelerographs, of which a large
variety are commercically available, have been derived
from the basic instrument of the 1930's.

The modern strong motion recorder is a rugged device designed to measure the ground motion in the epicentral (near-field) region of a strong earthquake. It
differs from the basic seismic instruments employed by
seismologists, which are too sensitive to operate in
the region of strong shaking. The unit contains three
transducers which are arranged to record simultaneously
the vertical and two orthogonal horizontal components
of ground acceleration as a function of time. The period and damping of the pick-ups are selected so that
the recorded motions are proportional to the ground acceleration over the frequency range of about 0.06 c.p.s.
to 25 c.p.s., which encompasses to the range of periods exhibited by typical engineering structures. The
instrument has a resolution of the order of 0.001 q
and is operative up to about 1.0 g. Records are normally taken on photo paper, 70 mm film or magnetic tape
at speeds of between 1-10 cm/sec., which are sufficient to provide for the accurate resolution of the
complicated wave forms for subsequent data analysis.
Since it is impractical to record continously, the
earthquake itself is used to trigger a vertical, horizontal or omnidirectional starter system, which in turn
activates the accelerograph. These switches have ad-
justable sensitivities and are normally pre-set to
operate at acceleration levels above 0.005 g so as to
avoid being set into motion by background noise (traffic, etc.); the recorder continues to operate for a
fixed period (7 sec.) after the motion has subsided be-

low this level. Starting times are of the order of
0.05-0.10 sec.; starting delays result in the loss of
the initial portions of records, although this does not
appear to be too significant. An independent power is
supplied by incorporating trickle-charge batteries in
the unit. The most recent models are essentially portable, weighing between 11-16 kg. and having outside
dimensions of the order of 20 x 20 x 40 inches. Their
starting prices are approximately $ 1350 U.S.

The specifications described above are not particular to one instrument, buth rather are an average compilation of the U.S. units. Suitable Japanese, New
Zealand and USSR recorders are also manufactured and
available for purchase. A summary of the characteristics of current strong motion accelerographs, which
has been prepared by Halverson,[1] is given in Table 1.

A growing appreciation of the type of fundamental
knowledge which can be derived from accelerograph records has resulted in a significant increase of the
number of instruments installed in the seismic areas of
the world during the last decade. The total number now
stands at about 2200 and it is likely that this figure
will be doubled in the reasonable future. Approximately 600 important records have been secured since the
program was introduced in the 1930's; copies of strong
motion data in a variety of forms are available on re-
quest from sources in the USA[2] and Japan.[3]

Seismoscopes

The data secured from strong motion accelerographs
can be inexpensively supplemented by means of a simple
instrument known as a seismoscope. This is a two dimensional pendulum with a defined period and damping,
whose total horizontal amplitude response to an earthquake excitation is scratched onto a smoked glass plate
without reference to time. In effect, a real structure
having the same dynamic characteristics as the pendulum
model would exhibit this same response when subjected
to this excitation. The maximum displacement of the
structure relative to the ground can be extracted from
the seismoscope record.

Seismoscopes are frequently arranged in clusters
to sample varying site conditions in the general area
(25-50 sq. miles) surrounding an accelerograph. They
offer a means of studying, at low cost, the manner in
which response levels appear to be influenced by local

Character-istics	USA USC & GS-Standard	Japan Akashi SMAC E	USA SMA-1	USA AR-240	New Zealand MO-2	USA RFT-250	USA RMT-280
Components							
Accelero-meters	2 horiz and 1 vertical	2 horiz and 1 vertical	2 horiz and 1 vertical	2 horiz and 1 vertical	2 horiz and 1 vertical	2 horiz and 1 vertical	2 horiz and 1 vertical
Displacement meters	2 horiz	None	None	None	None	None	None
Nat.per.(sec)							
Accelerom.	Adjustable (0.001-0.15)	0.05	0.05-0.07	0.04	0.03	0.05-0.08	0.05-0.08
Displacement meters	Adj.0.5-5.0sec (Carder)10-sec nominal(Coast Survey)	None	None	None	None	None	None
Sensitivity (in mm/o.1g)							
Acc.meters	Adjustable 6-20	0.5mag.-x8	1.9or3.8or 7-6	Adjustable 5.0-7.5	1.5horiz 2.2vertical	1.9	0.005
Damping(%of critical)							
Accelerom.	Adjustable 60%nominal	60%	Adjustable 60%nominal	Adjustable 55-65%	Adjustable	60%	60%
Displacement meters	Adjustable 60%nominal	None	None	None	None	None	None
Damping mechanism							
Accelerom.	Magnetic	Air piston	Electromagn.	Electromagn.	Viscous(oil)	Electrom.	Electrom.
Displacement meters	Magnetic	None	None	None	None	None	None
Recording range	0.001-1.0g (usually 1.0 in California)	±10-1,000gal	0.01-1.0g 0.005-0.5g or 0.025-0.25 g	0.01-1.0g	0.01g to 1.0 g	0.01-1.0g 0.005-0.5g 0.0025-0.025g	0.005-1.0g 0.0025-0.5 0.0012-0.25g
Traces and trace width	12 total 5 fixed-x,y,z and displ. 5variable-x, y,z and displ. 2 timing-each side Width:0.5mm to 0.75mm (depending on lamp current and develop-ment)	5 total 1 fixed 3 variable-x, y,z 1 timing Width:20lines/ mm	4 total 3 variable-x, y,z 1 timing Width:0.1mm	8 total 3fixed-x,y,z 3 variable-x, y,z 2 timing each side Width:0.25mm 0.38 mm	5 total 1 fixed 3 variable 1 timing Width: 0.075 mm on 35 mm film	7 total 3 fixed 3 variable 1 timing 0.1 mm	4 total 3 variable 1 timing reference track N/A
Time marking	2 per sec	5 per second	2 per second	2 per second ±1%	50 Hz and 5Hz	2 per second	2 per sec
Recording medium,speed and duration	Photo paper roll at 1cm/ sec(alternate drum) Duration: 1 1/4 min. and repeat-able for 5 cycles	Scratch record film,2.5mm/ sec Duration: 90 sec, able to start 100 times	70mm photo film at 1cm/ sec 15m total Duration:Adj. Nominal 10 sec. after last act. shock,and repeatable to end of roll	Photo paper roll,46m at 2 cm/sec Duration: 7 sec after last strong seismic shock and repeatable to end of paper roll	35 mm film 6 m total Duration: 47 s c at 1.5 cm/sec.Starts 9 times or 70 sec at 1.5cm/ sec starts 5 times	70mm photo film at 1cm/ sec 30 m to-tal Duration: 7 sec after last strong shock and repeatable to end of film roll	1/4" film cartridge Duration: 3 3/4 in/sec up to one hour
Recording drive	DC electric motor	DC electric motor(12vdc)	AC timing motor with solid state multibrator	DC governed electric motor (12 vdc)	Precision 12 vdc electric motor	DC governed electric motor (12 vdc)	DC electric motor (24 vol)
Built-in-cal-ibration sy-stem	None	Optional	Electrical impulse to produce a record of period and damping	Electrical impulse to record damp-ing and per-iod inform-ation	None	Electrical impulse to record damp-ing and per-iod inform-ation	Electrical impulse to record damping and period information
Starter com-ponent Type	Horiz.-closed relay Pendulum-el. contact	Vertical Pendulum-el. contact	Vertical Electromagn.	Horiz.-closed relay Pendulum-el. contact	Vertical Pendulum-emf (no contact)	Horiz.-open relay Pendulum-el. contact	Horiz.-open relay Pendulum-el. contact
Period Sensitivity	1 sec 0.05cm displ. of centre	0.3 sec 0.01g/0.005g	0.25 sec Adjustable 0.01g nom.	App 1 sec Adjustable to 0.01g	0-2 sec Velocity app.	App 1 sec	App 1 sec
Damping	30% critical -oil type	Not known	1.5 critical	100% critical -eddy current	Low	App.60%	App.60%
Starting time	App.0.2 sec	Not known	App.0.05 sec	App.0.1-0.15 sec	App.0.1 sec	App.0.1 sec	App.0.1sec
Power supply	12 vdc ex-ternal wet storage bat-teries with 115vac 60 cps ext.trickle charger	12 vdc in-ternal dry cells or al-kaline cells auto-charged	12 vds throw-away dry cells	12 vdc ex-ternal wet storage bat-teries with 115 vac in-ternal trickle charger	12 vdc ex-ternal re-chargeable batteries	12 vdc(2-6 V recharge-able lead -dioxide bat-teries)	24 vdc(4-6V recharge-able lead -dioxide batteries)
Size(cm) HxWxL	33x51x114	35x46x46 - incl.batte-ries	18x18x38	35x41x41	18x18x43	25x23x48	23x38x48
Weight(kg)	61 with cover	86 with steel cover	9 with cover	27 with alu-min. cover	9 less cover	16 with cover	19 with cover
Manufacturer or designer	USC&GS not commerci-ally avail-able.	Akashi Seisakusho Ltd Tokyo,Japan	Kine-Metrics Inc. San Gabriel, California	Teledyne/ Geotech. Garland,Texas	Victoria,Eng. Ltd.Naenae, N.Z.Repr. worldwide by Teledyne/Geo-tech.	Teledyne/ Geotech. Garland,Texas	Teledyne/ Geotech. Garland, Texas.
Present price							
Other accelerographs	-	Ishimoto SMAC-A,B,C, D, and Akashi IAS-P acceler-ographs	SMA-2,SMA-3	RFT-250 RMT-280 MO-2,PRA,100, FB-100		RMT-280 MO-2, PRA-100 FB-100	RFT-250 MO-2, PRA-100, FB-100

Table 1: Characteristics of significant strong-motion accelerographs (Based on best information available Dec. 1969). (Ref 1)

geology and soil conditions. The correlation between
the system response calculated from accelerogram data
and the actual response measured with a seismoscope has
been found to be quite satisfactory.[4]

Excellent summaries of strong motion instrumentation have been published by Hudson[5] and Halverson[6].
Photographs of an accelerograph and seismoscope are
reproduced in Figure 1.

CHARACTERISTICS OF STRONG MOTION RECORDS

Accelerograms

Having discussed the need for and method of securing strong motion records, the characteristics of an
actual accelerogram will now be examined. Figure 2 reproduces typical acceleration traces obtained from an
area subjected to strong shaking due to destructive
earthquakes (El Centro, 1940 and Taft, 1952). Their
complexity suggests a random process involving a wide
range of frequencies. There is a rapid, initial rise
to a strong and relatively uniform central phase of
shaking for a certain duration followed by a gradually
decaying tail during which some very strong pulses of
acceleration may still occur. The records exhibit a-
bout equal intensities of motion in the two horizontal
directions, which is to be expected because of the almost random nature of the ground movement. The vertical component is normally somewhat less intense than
the horizontal and is characterized by an accentuation
of the higher frequency components when compared with
the horizontal motions.

Ground velocities and displacements can be calculated directly from the acceleration records by integration procedures. Representative curves (El Centro
1940 NS) of these motions are illustrated in Figure 3.
The appearance of the plots is different for each parameter and the frequency regimes associated with the
different motions are emphasized. The acceleration
record exhibits many high frequency components and the
larger accelerations are associated with these frequencies. The displacement motion is dominated by the low
frequency components while the velocity samples the
intermediate frequency range. Although the acceleration is generally the most important parameter, the
particle velocity and displacement also provide useful
information. The latter quantity is indicative of the

Figure 1: Strong motion recorders

94

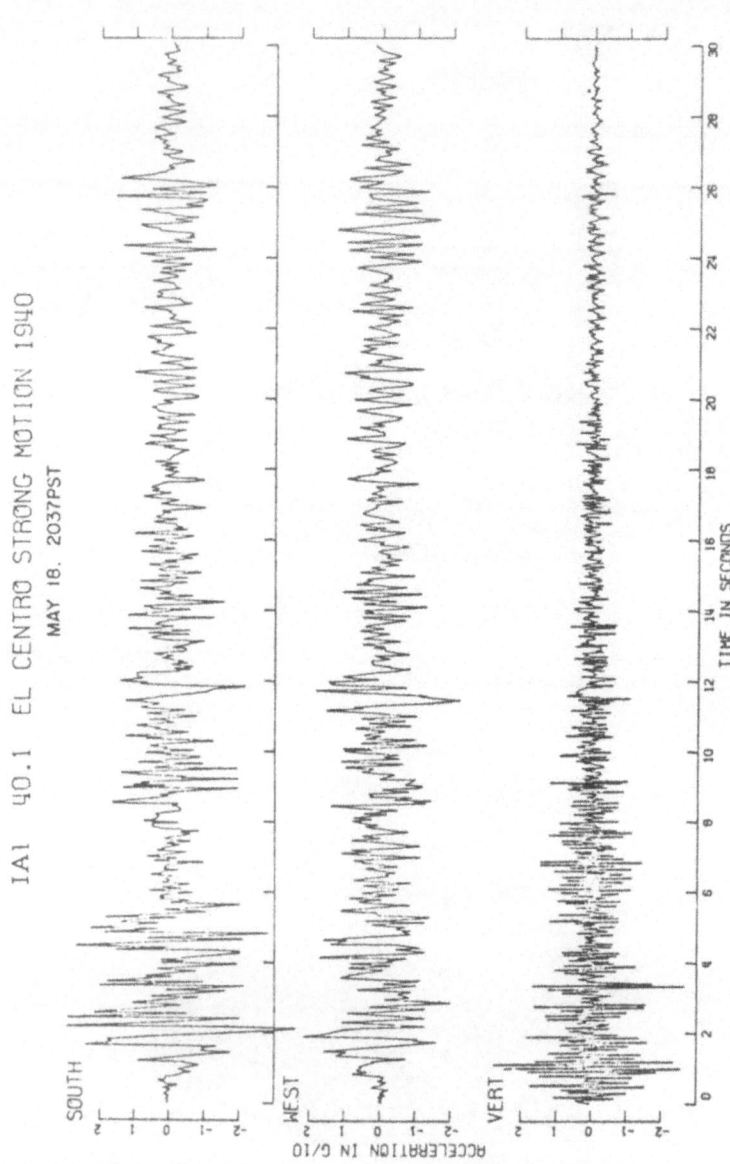

Figure 2a: Strong motion accelerograms

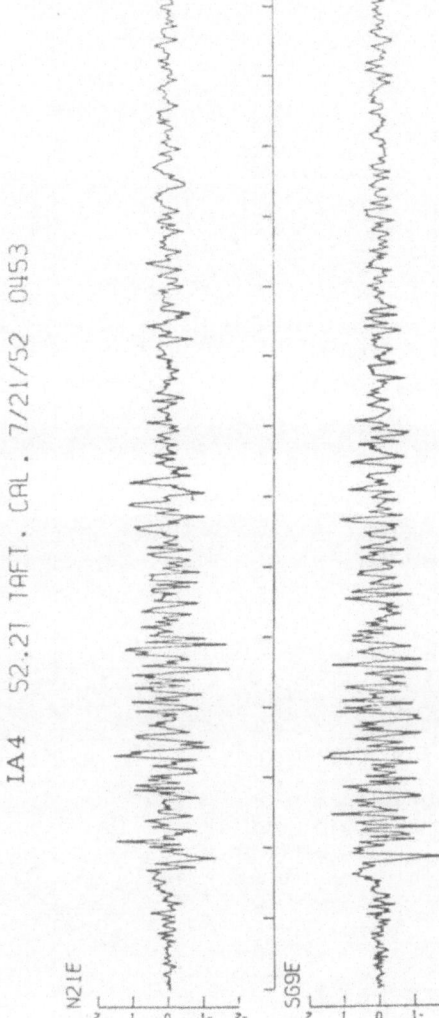

Figure 2b: Strong motion accelerograms

96

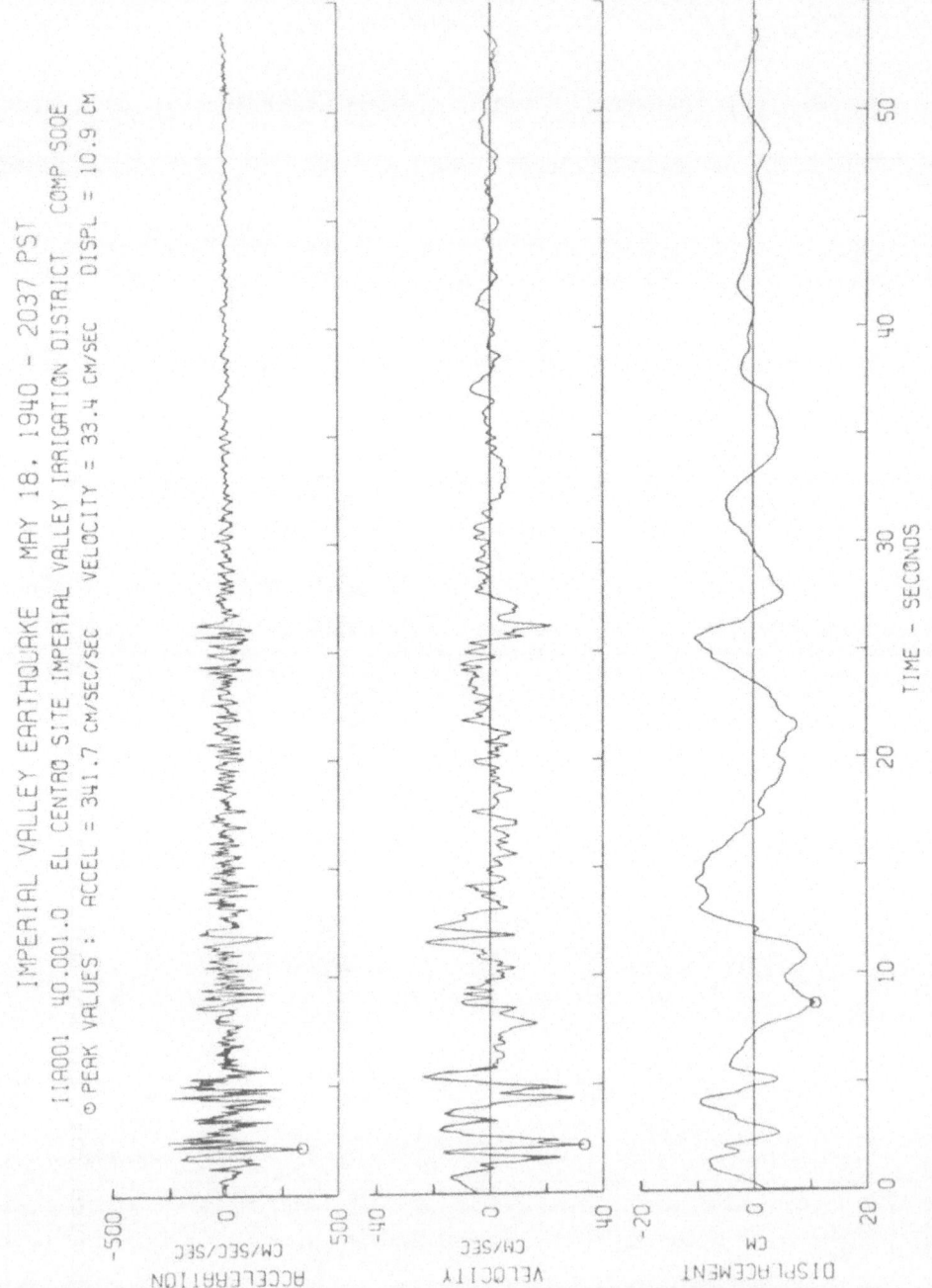

Figure 3: Recorded acceleration and integrated velocity and displacement

strains to which large structures and underground pip-
ing systems may be subjected, while the former parameter is directly related to the input energy and as
such approximately correlates with intensity of structural damage.

Fourier Spectrum

The frequency content of an acceleration signal
can be characterized by the Fourier amplitude spectrum,[*]
which is obtained by a mathematical exercise that
transforms the time function into the frequency domain.
This process, illustrated diagrammatically in Figure 4,
represents a convenient means of identifying the predominant frequencies in the record, which are the ones
associated with relatively large proportions of the total energy content in the input signal. Such curves
are indicative of the relative manner in which a system
of structures will respond to the excitation function.
The Fourier spectrum.plot for the Taft accelerogram of
Figure 2 is shown in Figure 5. The graph indicates
that the maximum accelerations were basically due to
frequencies in the 1/4 to 7 c.p.s. range; this hap-
pens to fall within the frequency region exhibited by
many engineering structures.

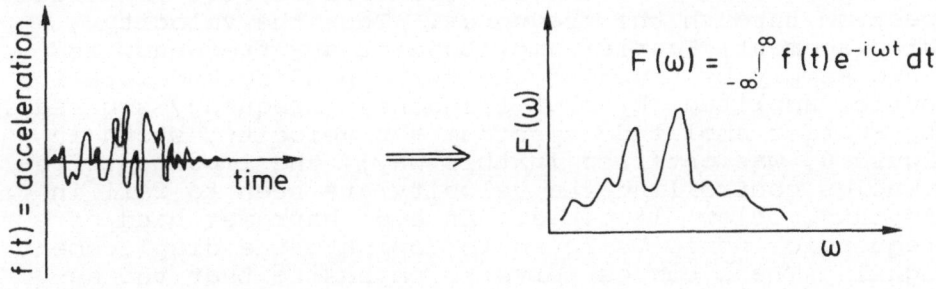

$$F(\omega) = \int_{-\infty}^{\infty} f(t)e^{-i\omega t}\,dt$$

Figure 4:   Fourier spectrum for earthquake ground
            motion.

---

* A related concept is the power spectral density
  function which expresses the relative value of the
  energy content at each frequency. This function is
  proportional to the square of the Fourier amplitude
  spectrum.

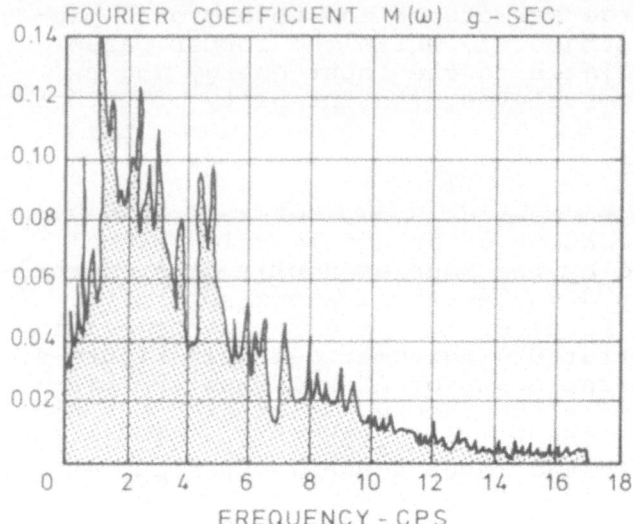

Figure 5: Fourier acceleration spectrum of Taft earthquake (Ref.7).

The Fourier spectrum for velocity (displacement)
is simply related to the acceleration Fourier amplitude
spectrum through the frequency. Thus the velocity
(displacement) Fourier amplitude at any frequency is
found simply by dividing the corresponding acceleration
Fourier amplitude by that frequency (frequency squared).
The Fourier amplitude spectrum for velocity, shown in
Figure 6, was obtained in this way; the important frequencies controlling the velocity are seen to fall in
the range below 1½ c.p.s. An even narrower band of
frequencies would be found to dominate the displacement
signal. These curves serve to emphasize that the acceleration, velocity and displacements are controlled
by different frequency bands within the total frequency
regime of a signal.

Response Spectrum

The information contained in a ground motion record can also be examined by constructing a structural
response spectrum. This concept has a special appeal
for earthquake engineers since it combines the significant features of the ground excitation (amplitude,
frequency content and duration of shaking) and of the
structure (natural period and damping) in the form of a

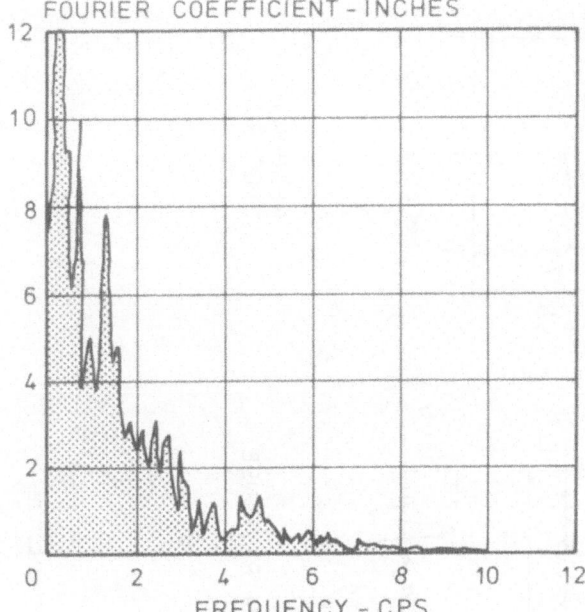

Figure 6:   Velocity Fourier spectrum of Taft record
            (Ref.7)

structural response calculation. The response spectrum
is defined as the maximum response of a linear, single
degree-of-freedom structure to a specific ground acceleration, plotted as a function of the natural period or
frequency and damping of the structure. A schematic
interpretation of the quantity is shown in Figure 7.
Once the spectrum for an earthquake is known it is possible to compute the forces developed by that earthquake. The response may be expressed in terms of the
displacement, velocity or acceleration of the structure.
These quantities are approximately related to each
other by the natural frequency of the system in the
manner noted above for the Fourier spectrum relationship.

Examples of separate acceleration and velocity response spectra are given in Figure 8 (El Centro). Because of the simple relationship between the three response parameters (peak acceleration, velocity and displacement), their spectra can be plotted on a single
tripartite graph, as shown in Figure 9 (Taft). The
major spikes in the lightly damped curves indicate that
structures having natural periods which coincide with
these peak locations would be strongly excited by the
particular earthquake in question. The peaks represent

100

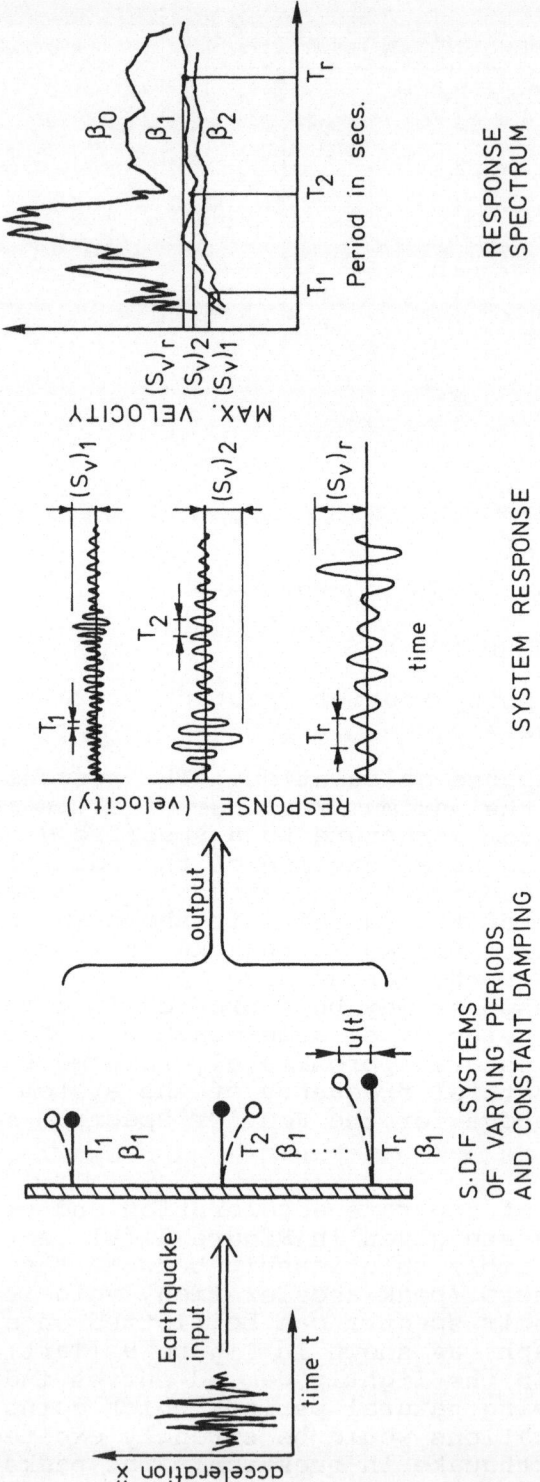

Figure 7: Scematic interpretation of response spectrum

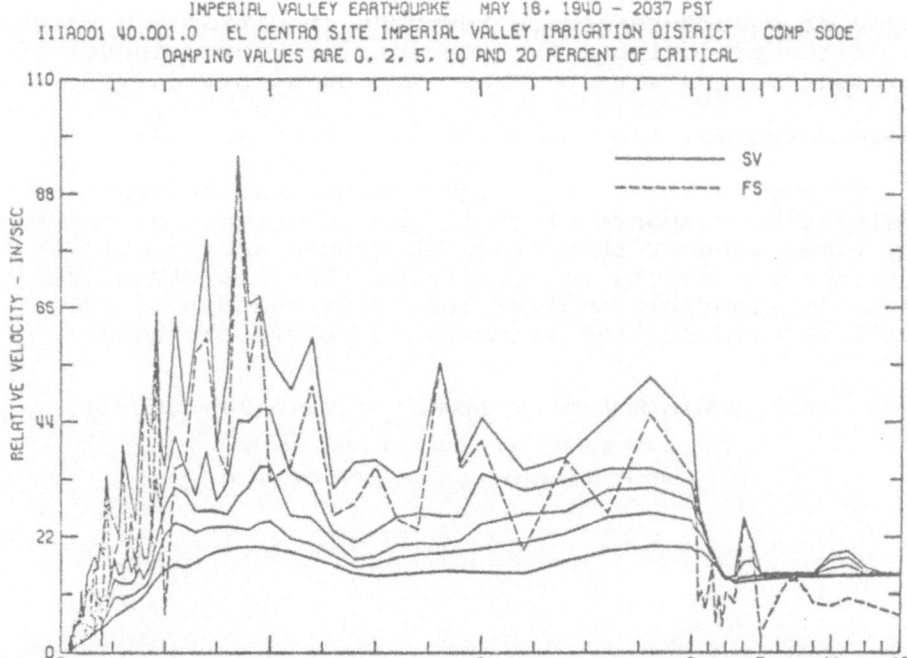

Figure 8a: Relative velocity response spectrum of El
Centro, NS, 1940.

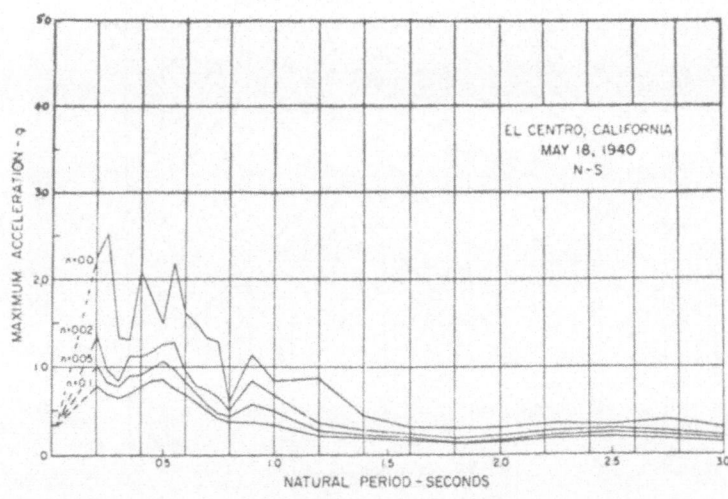

Figure 8b: Acceleration response spectrum of El Centro,
NS, 1940.

a type of pseudo-resonance response condition - a matching of the predominant frequencies (or energy input) of the gound motion with natural structural periods.

Damage Potential

It may be noted from Figure 8(b) that the maximum acceleration response of short period structures may be many times greater than the peak ground acceleration, while the corresponding values for flexible structures may be considerably smaller than this quantity. It is therefore evident that attempts to relate earthquake

KERN COUNTY, CALIFORNIA EARTHQUAKE   JULY 21, 1952 - 0453 PDT

IIIA004 52.002.0   TAFT LINCOLN SCHOOL TUNNEL   COMP S69E

DAMPING VALUES ARE 0, 2, 5, 10 AND 20 PERCENT OF CRITICAL

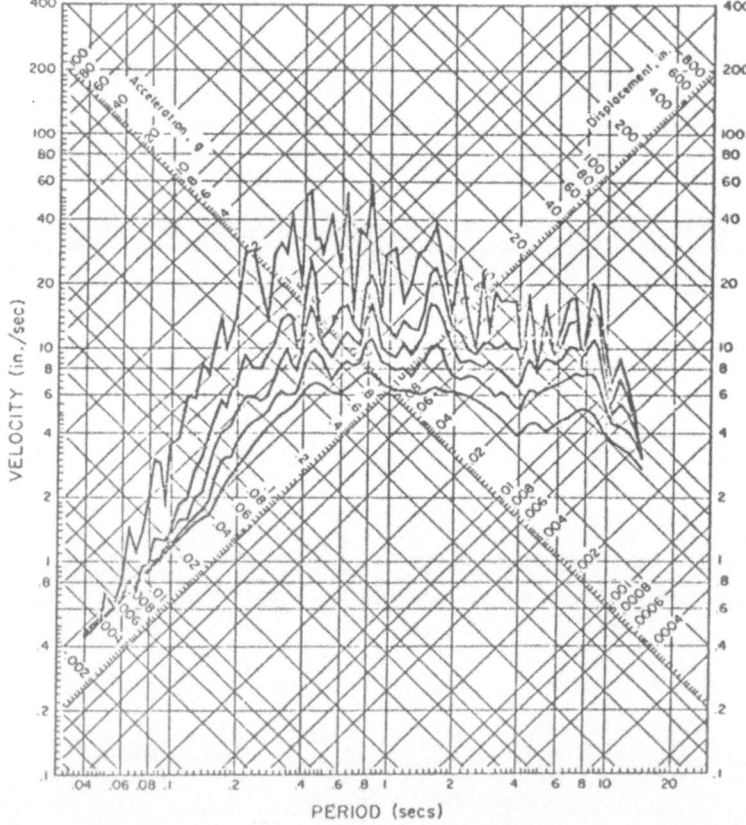

Figure 9:   Tripartite response spectrum of Taft, S69E, 1952.

damage potential to a single parameter, such as peak
ground acceleration, which is intended as a measure of
seismic force levels reached, are normally unsatis-
factory and inappropriate. Such relationships fail to
account for the interaction of the characteristics of
the structure and the characteristics of the acceleration-time history, including its duration of shaking
and the area and number of acceleration pulses in the
record. The duration of the acceleration above various
levels is a more useful guide for estimating damage potential than the maximum ground acceleration. A more
sophisticated type of response spectrum, Figure 10,
showing the length of and position in time different
structures experience certain levels of peak velocity,
which has recently been suggested by Perez,[8] reflects
this concept.

A single numerical quantity based on instrumental
records and commonly employed by engineers as a quantitative measure of earthquake damage potential is the
Spectrum Intensity, which was introduced by Housner.[9]
It is defined as the area under the velocity response
spectrum over a specified period range and for a standard damping value, Figure 11, and is a measure of the
intensity of ground shaking in the sense that it expresses the average response of elastic structures to
this shaking. For any earthquake, the spectrum intensity is an indication of the energy that structure on
the average must be able to absorb or dissipate if
damage is to be avoided.

FACTORS AFFECTING ACCELEROGRAM CHARACTERISTICS

As noted above, methods of examining and expressing the characteristics of actual records are readily
available. However, the designer is still faced with
the uncertainty of deciding on the nature of the appropriate earthquake expected at a particular site. The
determination of the seismic exposure of a site involves a study of historical, seismological (instrumental), geological and geophysical data. From such
information it is possible to estimate the potential
location, magnitude and frequency of occurrence of seismic events, the type, length and amount of faulting,
the duration of shaking and the focal depth and distance to causative fault. These details control the
basic ground motion parameters (acceleration amplitude,
frequency content and duration of strong shaking) which
should be exhibited by the design earthquake. Acceleration time-history records reflecting the desired pro-

104

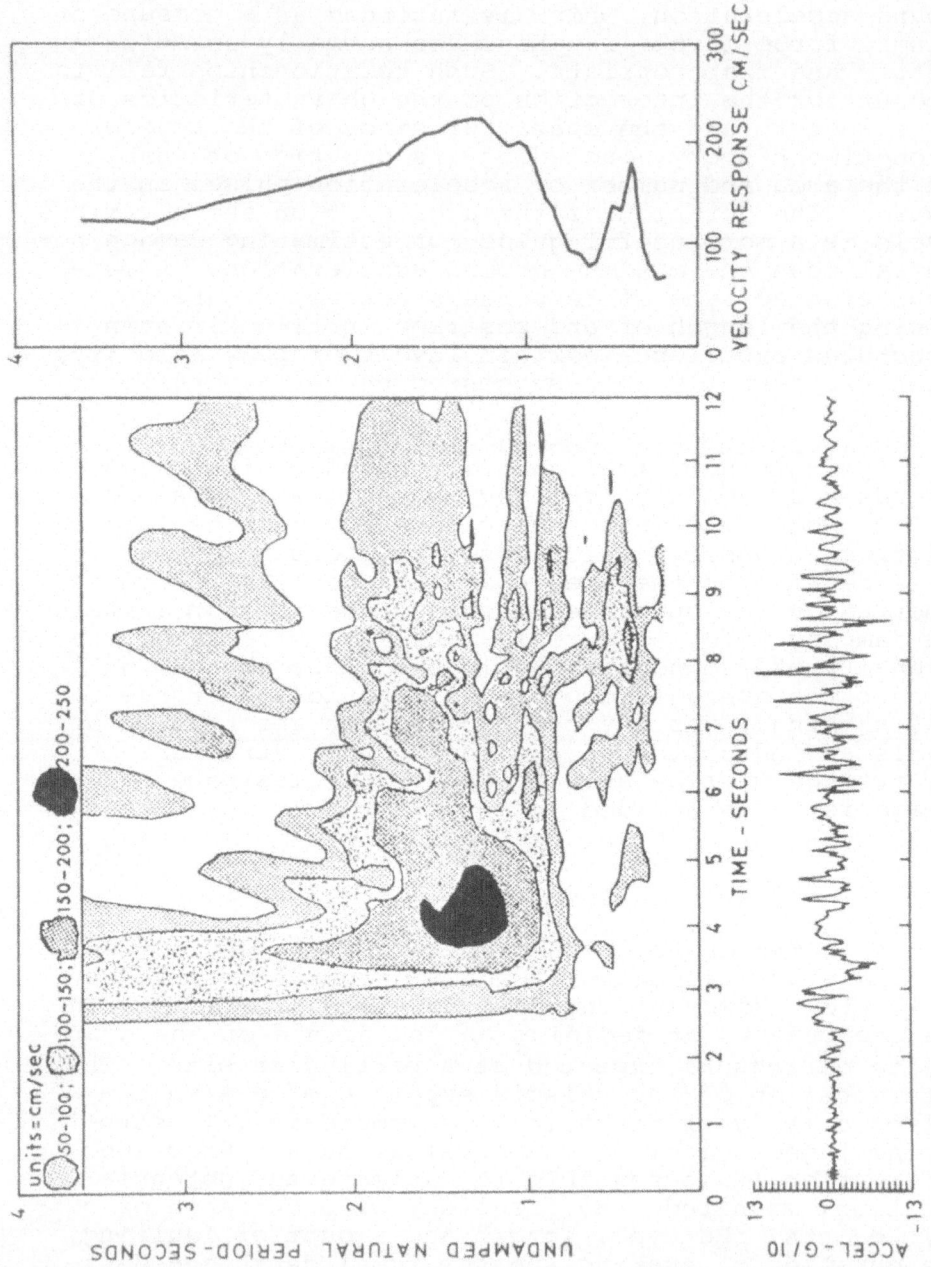

Figure 10: Velocity response envelope spectrum (VRES), 5 per cent critical damping for Pacoima Dam, S16°E direction. (Ref.8).

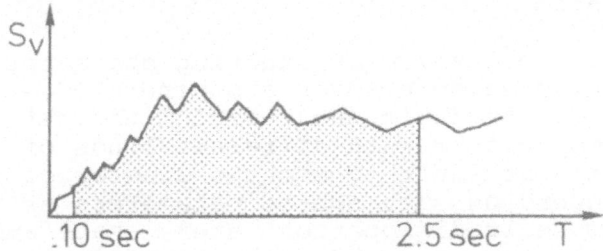

Figure 11:   Definition of Housner's Spectral Intensity

perties may be derived by (i) direct extrapolation
from comparable recorded accelerograms (ii) modification of existing records (iii) use of artificially constructed earthquakes based on the theory of stochastic
processes.

Since existing accelerograms are frequently used
by engineers to describe the signature of future events,
the factors which may influence and complicate the surface ground motions measured with an accelerograph
should be appreciated before deciding on the choice of
an appropriate catalogue record for response calculations.   These factors are illustrated diagrammatically
in Figure 12 and include the nature of the source mechanism, the transmission path geology and the local site
conditions.   They are discussed briefly below and our
current knowledge of their effects on measured ground
motions is also illustrated.

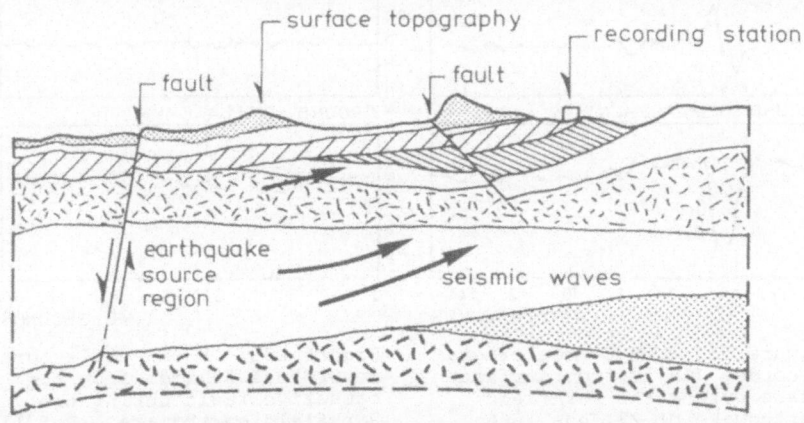

Figure 12:   Schematic diagram showing relation of earthquake source, travel path, and local site

106

Source Mechanism

The source mechanism or faulting character is the process by which seismic waves are generated, and in the vicinity of the fault (near-field) the nature of these waves may differ with different kinds of faults. For any particular fault type the character of these waves will depend on such source parameters as stress drop, maximum fault dislocation, area, shape and nature of fault surface, and relation of fault plane to ground surface. The near-field peak acceleration, frequency content and duration of shaking are mainly functions of the dislocation on the faulting surface. However, these ground motion parameters are altered as the wave trains propagate from the fault to the site such that at distances of several miles from the causative fault the actual fault type does not significantly influence the strong phase of a surface acceleration record. Figure 13 illustrates an example of the change in overall character of the ground motions with length of wave travel; some few miles from the fault the shaking appears as a random process while the measurements adjacent to the fault reflect the displacement history of that phenomenon.

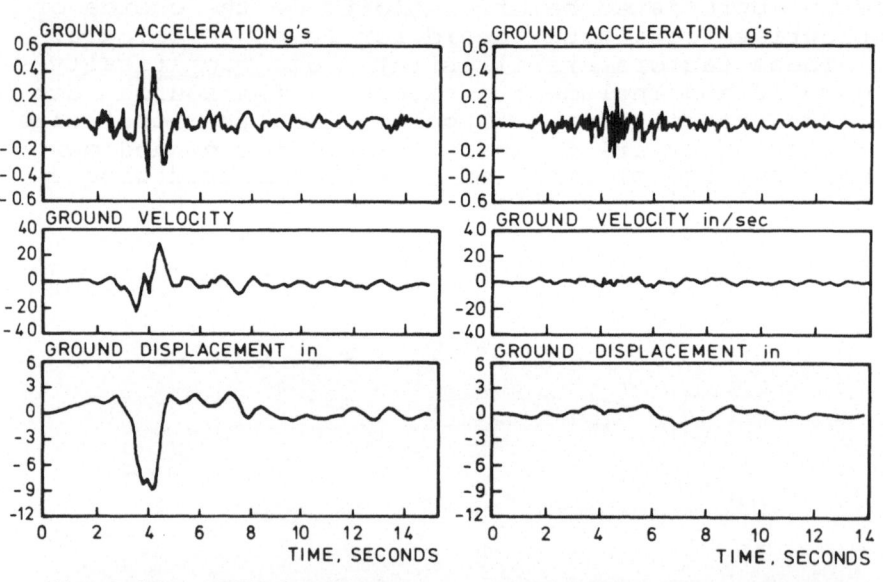

Figure 13a: N65E accelerogram recorded 200 ft from surface trace during the Parkfield earthquake of 27.June 1966; magnitude 5.6; strikeslip fault. (Ref.11).

Figure 13b: N50E accelerogram recorded 5.3 miles from the causative fault during the Parkfield earthquake. (Ref.11).

Transmission Path

Modification of the ground motion parameters initiated at the source, such as noted above, are due to
the geological structure along the transmission path
and the site conditions, particularly the surficial
layers. Although geologic structure may enhance local
shaking because of energy focussing effects, there is
normally an attenuation or reduction of acceleration
amplitude as the waves travel from the causative fault
to the measuring site. This is attributed to decreases
in the specific seismic energy, which occur due to geometric and material dispersion as the outwardly propa-
gating waves occupy a larger volume, and to the energy
absorption associated with internal damping in the rock
material. Examples of measured peak acceleration-at-
tenuation curves which have been reported by Cloud and
Perez [12] and by Hudson [13] are shown in Figures 14 and
15. The scatter in values for the same event reflect
the influence of travel path and local geologic environment.

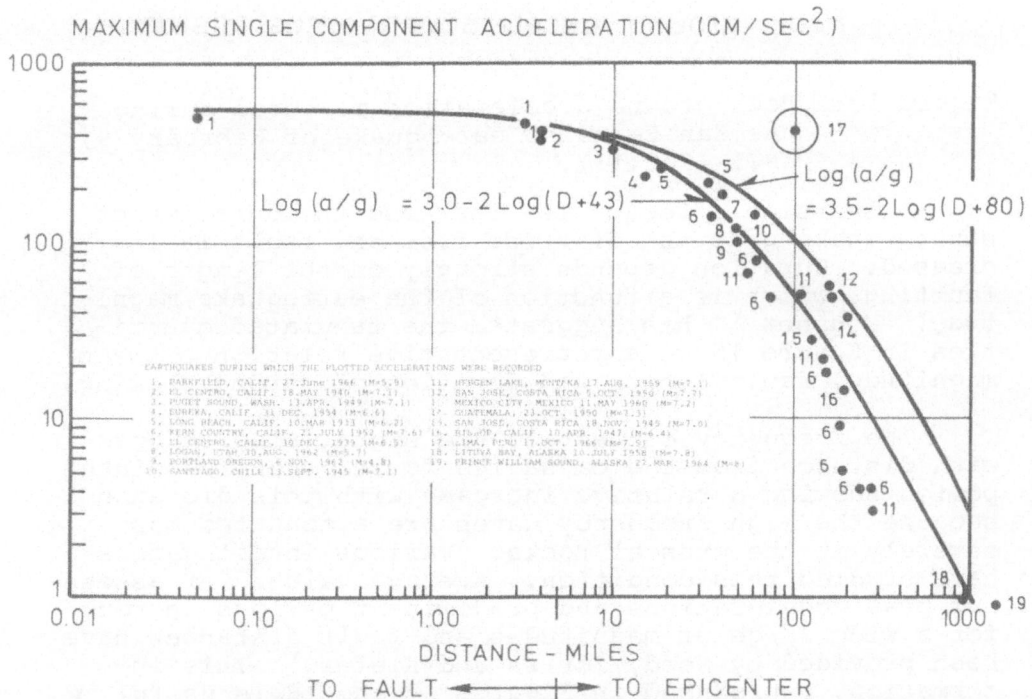

Figure 14: Attenuation of maximum acceleration (Ref.12)

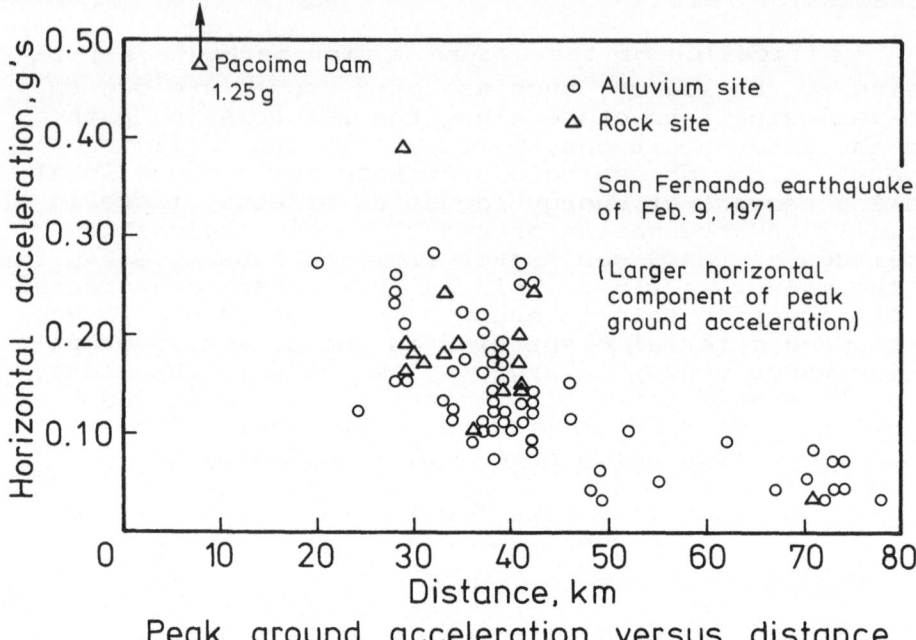

Peak ground acceleration versus distance

Figure 15: Peak ground acceleration measured during
the San Fernando earthquake of February 9,
1971 (Ref.13).

Dispersion effects also increase the duration of
strong shaking as the distance from the fault is increased. Duration depends strongly on the length of
faulting, which is a function of the earthquake magnitude. Housner [14] has suggested the tabulated quantities in Figure 16 as a representative relation between
magnitude, fault length and duration of strong shaking.

The frequency content of the signal also changes
with distance from the causative fault, the predominant
period showing a relative increase with this distance
because the high frequency waves are attenuated most
severely in the crustal rocks. Various investigators
have studied this condition. Average values for assessing peak accelerations and predominant periods in rock
for a wide range of magnitudes and fault distances have
been provided by Seed, Idriss and Kiefer.[15] This information, reproduced in Figures 17 and 18 is useful for
seismic site evaluation and the extrapolation and
scaling of recorded accelerograms for design earthquake
ground motions.

Magnitude of Earthquake (M)	Approx. Length of Fault Slip (miles)	Approx. Duration of Strong Shaking (secs.)
3	0.3	
4	0.8	
5	2	1
6	5	8
7	25	20
8	190	35
8.5	530	

Figure 16: Relation between magnitude, fault length and duration of shaking, (Ref.14).

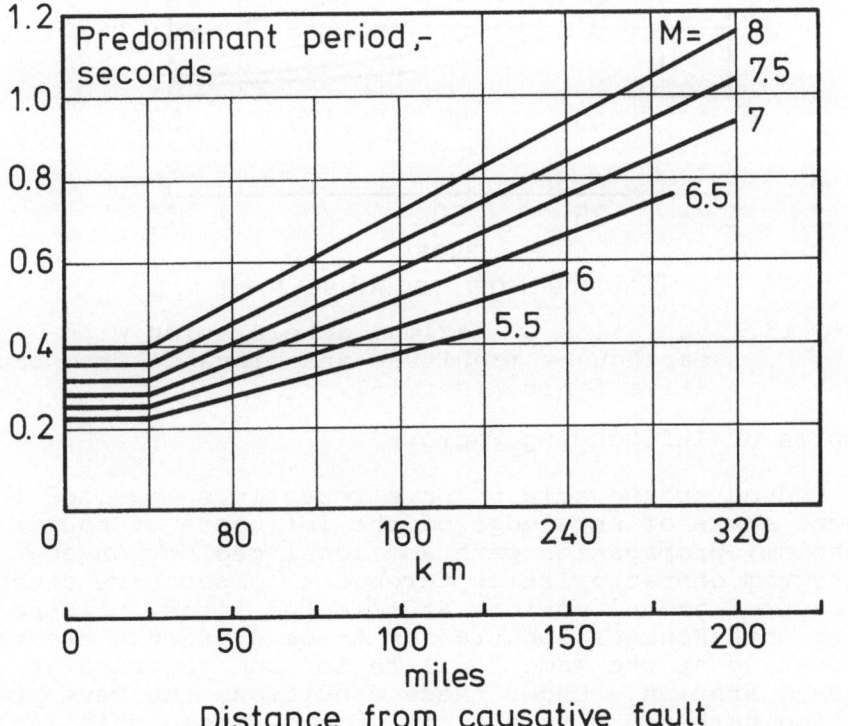

Figure 17: Predominant Periods for maximum accelerations in rock, (Ref.15).

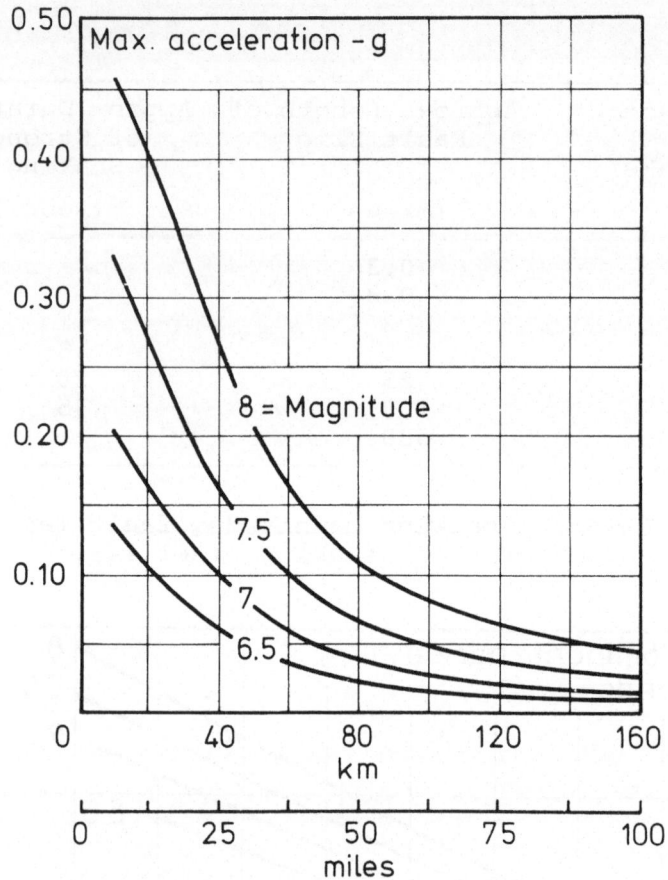

Figure 18:   Variation of maximum acceleration with
             earthquake magnitude and distance from caus-
             ative fault, (Ref.15).

Examples of Influencing Factors

        Hudson and Udwadia [16] have recently summarized the
current state of knowledge of the influence of source
mechanism, propagation path and local geology on accelerogram characteristics through a comparative study
of measured ground motions at selected sites. Figure
19 presents Fourier spectra for three different events
originating at the same focal region and recorded at
the same station. Under these conditions the wave propagation path and site effects should be essentially
identical so that the significant differences noted in
the spectra can be attributed to the influence of
source mechanism. The smaller events appear to be relatively richer in high frequency content and peaking
coincidence is non-existent.

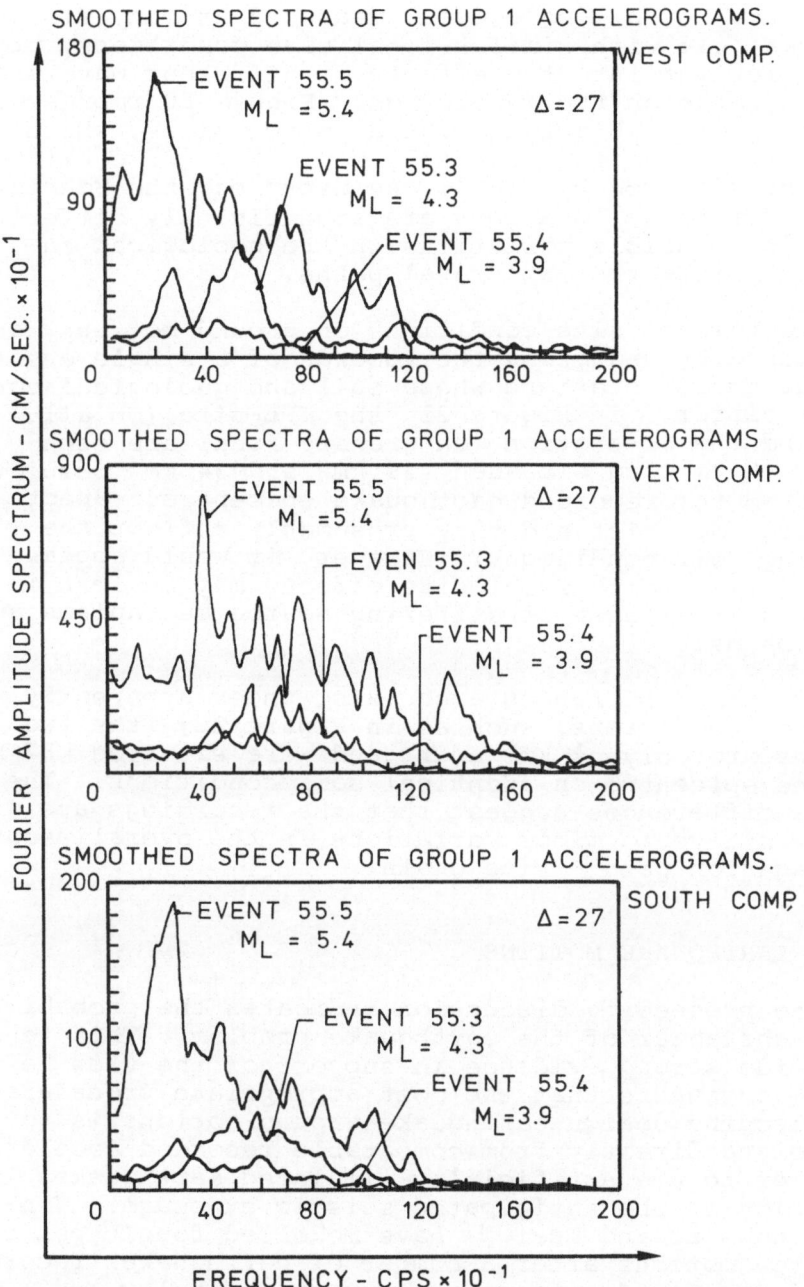

Figure 19: Fourier amplitude spectra calculated from accelerograms from the El Centro strong motion station, California. Events of December 16, 1955, (Ref.16).

Spectra of a single event measured at many different sites, all with similar local site conditions, provide an insight into the effects of different wave propagation paths on the recordings. Figure 20 represents an example of such a study based on two sites, whose epicentral distances are nearly equal but whose azimuth locations differed by 100^0. As expected, the Fourier spectra for these locations are significantly different, reflecting possible variations in the geological environment along the two travel paths.

The role of site conditions on ground motions can be examined by analyzing the records of a single event taken at nearby stations whose soil and geological conditions differ. In Figure 21 the JPL-site (on alluvium) and the SL-station (on granite rock) are separated by 6 km and situated 29 km and 34 km respectively from the recorded earthquake epicentre. Spectra variations do exist and they presumably reflect the differing soil conditions. However, it would appear that for firm soils local site effects may be less important than effects of differing source mechanisms and travel paths.

Spectra differences even exist under apparently identical conditions, such as in Figure 22; the two stations are only ½ km apart and were situated 37 km from the epicentre on identical soil conditions. The spectra differences suggest that the recordings are very sensitive to minor variations in the overall source - propagation path - site chain.

DESIGN EARTHQUAKE MOTIONS

The preceeding discussion indicates the probabilistic character of the earthquake problem. This tends to provide strong evidence in support of the view held by some engineers that the most appropriate procedure for selecting design earthquake ground motions is to extrapolate directly from comparable recorded accelerograms, or to use artificially simulated earthquakes conforming to the anticipated seismic exposure. Jennings, Housner and Tsai [17] have modelled four types of simulated motions after a number of earthquakes recorded on firm ground; these are summarized below and their Type B record is shown in Figure 23. Newmark and Rosenbleuth [18] have classified earthquakes into groups from a sampling of typical accelerograms, and these are correlated here, where possible, with the simulated types by offering examples of these motions.

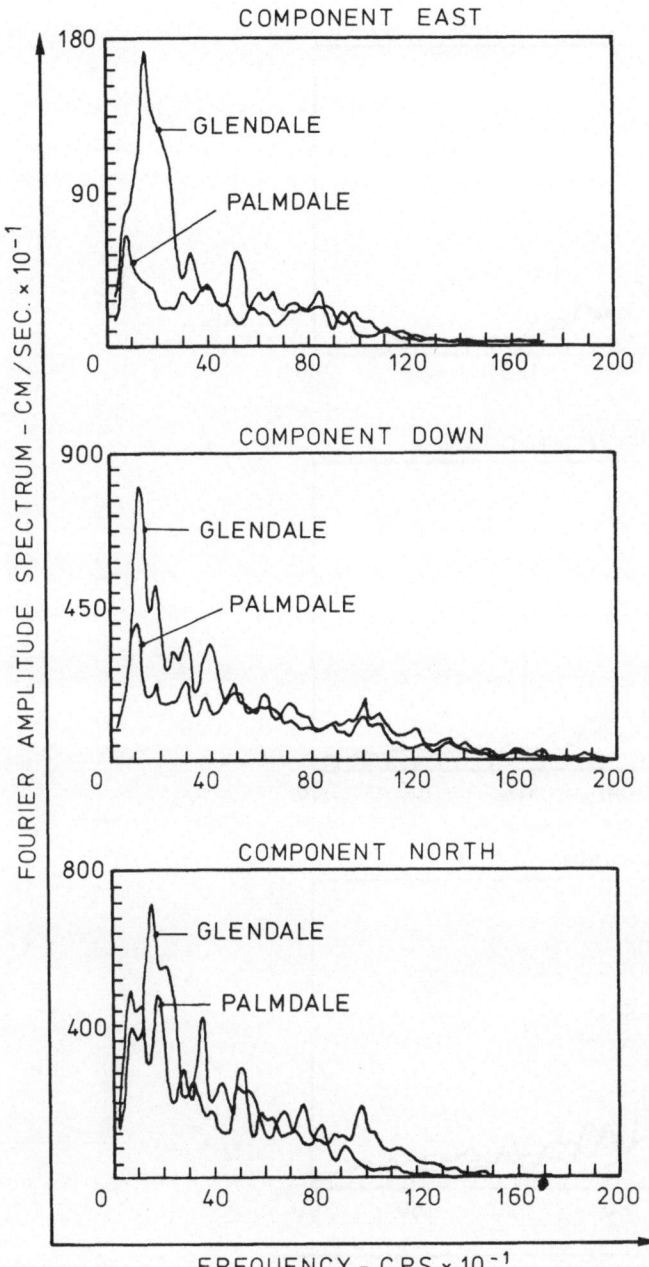

Figure 20: Fourier amplitude spectra calculated from accelerograms of the San Fernando earthquake of February 9, 1971, (Ref.16).

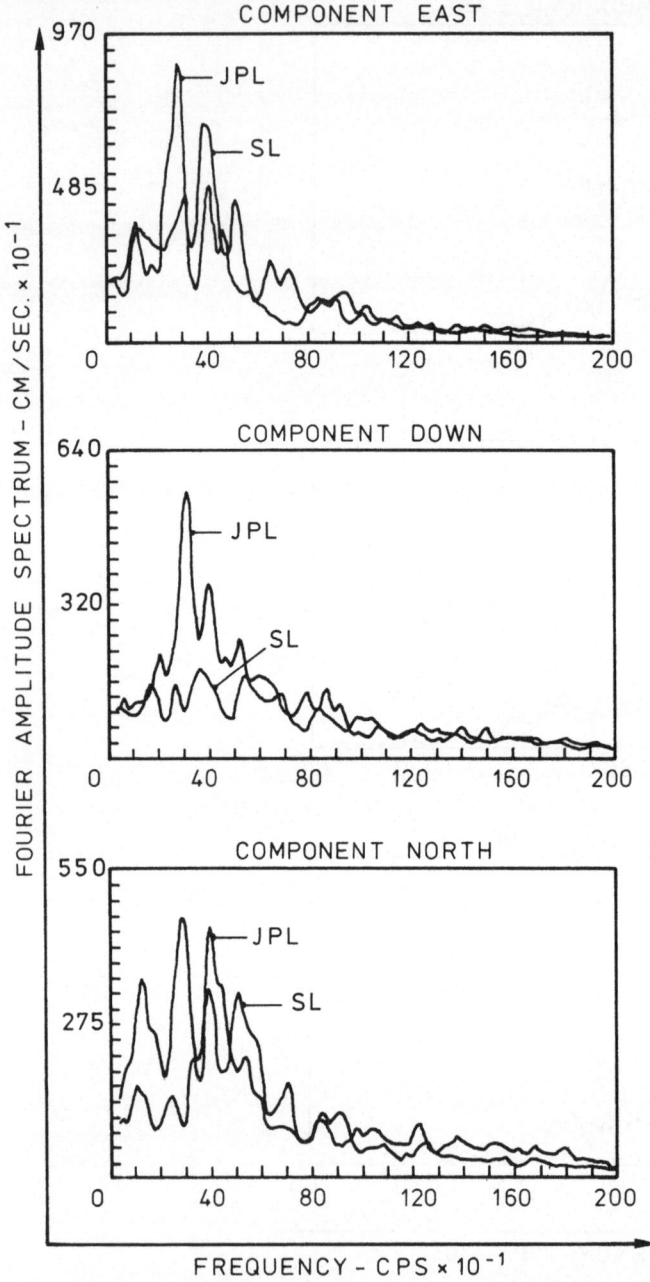

Figure 21: Fourier amplitude spectra calculated from accelerograms of the San Fernando earthquake of February 9, 1971

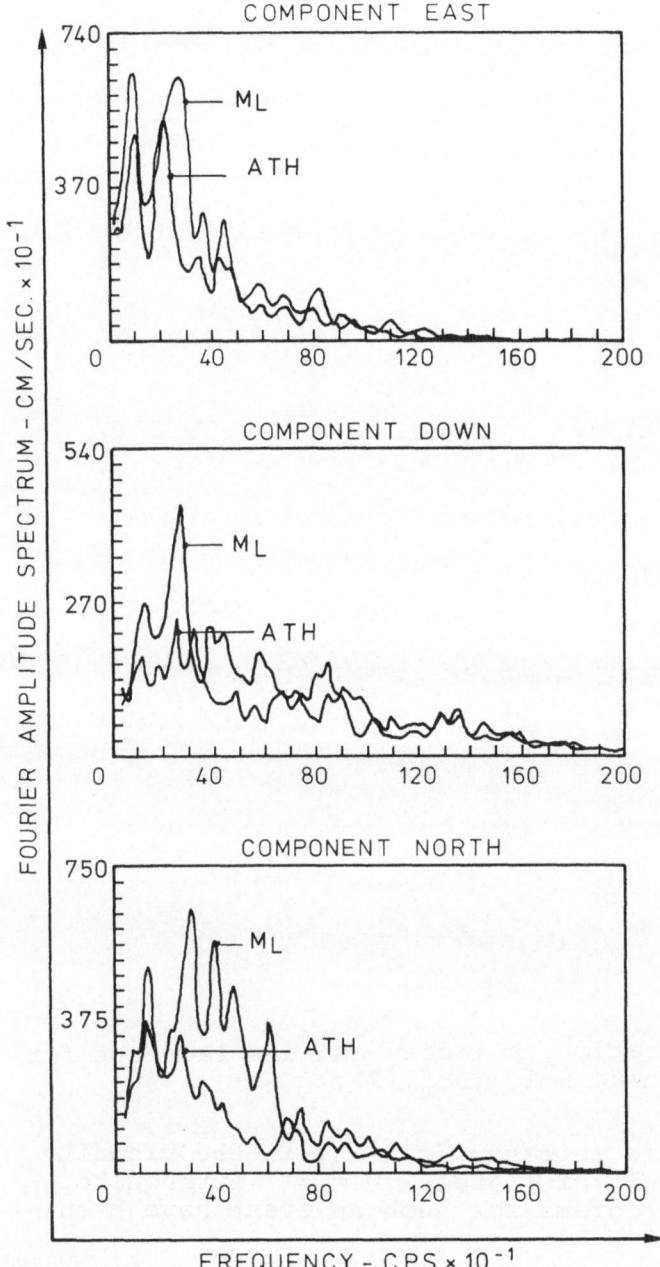

Figure 22:   Fourier amplitude spectra calculated from
             accelerograms of the San Fernando earth-
             quake of February 9, 1971.

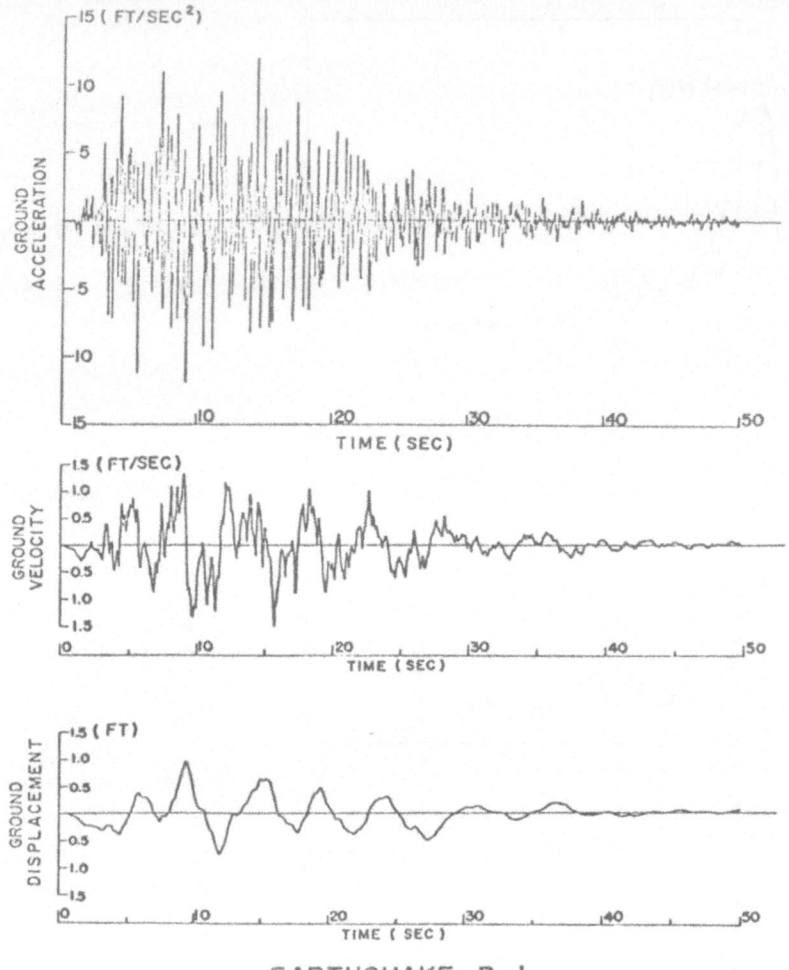

EARTHQUAKE B-I

Figure 23: Acceleration, velocity and displacement for earthquake B-1, (Ref.17).

Type A: Intended to represent shaking in the vicinity of the fault of a great M = 8+ earthquake. No accelerograms for such an event have been recorded.

Type B: Strong shaking associated with a M = 7 or 7+ earthquake exhibiting long, extremely irregular motion. Typical of motions on firm deep ground at moderate distances from the focus, e.g. El Centro NS 1940, M = 7.1 , 4 miles from causative fault.

Type C: Corresponds to motions that would probably be
          obtained from a recorder located on rock for a
          moderate M = 5.5-6 earthquake with short
          epicentral distance, e.g. 1957 San Fransisco
          Golden Gate, M = 5.3, 6.5 mile epicentral
          distance.

Type D: Records characterized by practically a single
          shock which corresponds to accelerograms ob-
          tained on firm ground close to the causative
          fault of shallow M = 4.5-6.5 earthquake, e.g.
          Parkfield 1966, M = 5.6, 200 ft. from surface
          trace.

Modified Earthquake Records

     Still another technique for selecting a design
earthquake has been pursued by Seed et al.[15]   Bedrock
motions underlying the site are first estimated by modifying an existing record, through the application of
appropriate scale factors, to provide the peak acceleration and predominant period consistent with the magnitude and focal distance for the anticipated earthquake.
The soil overburden is then idealized as a lumped mass
model and its time-history response to the rock motion
is calculated. The same approach has been used to predict the motions at one site from the ground measurements recorded at another and to estimate subsurface
motions from surface accelerograms. These concepts are
illustrated graphically in Figure 24. The assumptions

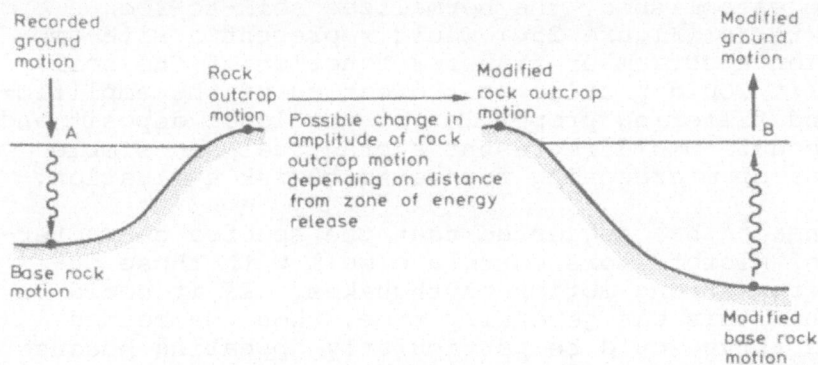

Figure 24:   Schematic representation of procedure for
             computing effects of local soil conditions
             on ground motions, (Ref.19).

involved in this approach are considered by some engineers to be a gross oversimplification of reality.

MICROTERMOR SPECTRA

Microtremors are continous ground motions whose
amplitudes range from 10^{-6} to 10^{-7} metres and whose
frequencies may range from ½ to 20+ Hz . They are
believed to originate primarily from artificial (man
made) sources, such as traffic, machinery etc. and uncontrolled natural sources such as wind, waves etc.
Their connection with this discussion stems from the
fact that attempts have been made to use such disturb-
ances to predict the seismic response characteristics
of a potential building site, or to assess the relative
response of different sites in the same region.

In the procedure, the recorded data are analyzed
by Fourier spectrum techniques in order to identify the
existence of any predominant site frequencies, which
might be attributed to the dynamic amplification characteristics of the local subsoil conditions. The
measured amplitudes of adjacent sites are also correlated relative to geology and soil type. Figure 25 [20]
represents the results of a typical microtremor survey.
The spectrum for the bedrock recordings (Figure 25a) is
reasonably flat or of "white noise" type, while the
spectrum for the 22 ft. layer of sandy, clayey material overlying bedrock (Figure 25b) is characterized by
a major spike. If the validity of the microtremor approach for establishing site response characteristics
could be established, the normalized soil-to-rock
spectra ratio (Figure 25c) would represent a site amplification spectrum or transfer function of the soil
layer; it would provide an indication of the amplification and filtering properties of the local deposit and
the procedure would represent a practical and simple
technique of microzoning for seismic risk evaluation.

Kanai [21] has suggested that the spectra characteristics of microtremors correlate well with those obtained from strong motion earthquakes. If it could be
shown that this was generally true, then the method
outlined above would be particularly appealing because
of the speed, ease and relatively low cost of securing
microtremor measurements. Despite some apparent correlations, [22] for reasons noted below this approach
does not appear to be reliable as a tool for microzonation.

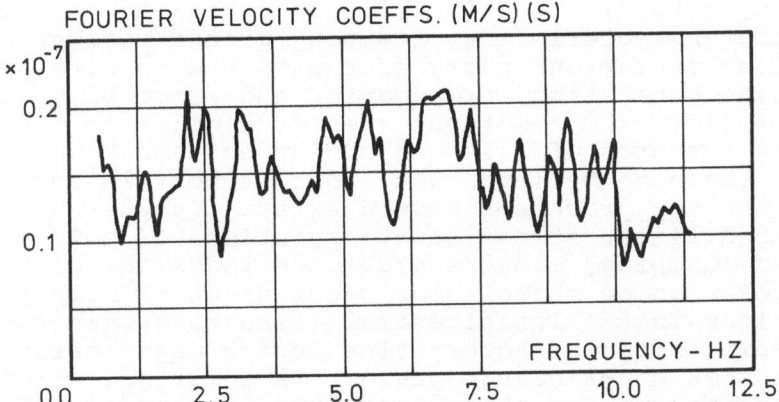

Figure 25a: Fourier Analysis of M7EH (Rock Outcrop
              500'E. Of M.Jenkins SCH 1).

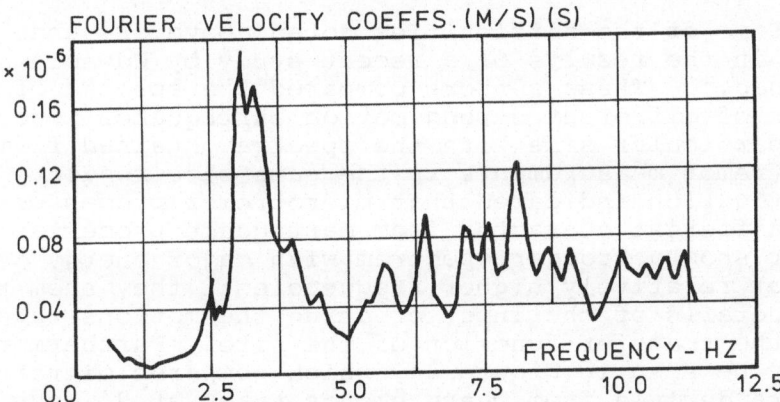

Figure 25b: Fourier Analysis of M7ES. (M.Jenkins SCH:)

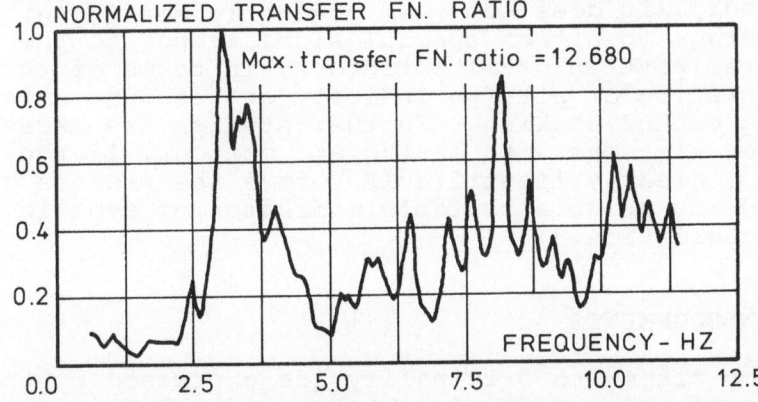

Figure 25c: Ratio of Vel. Spectra(EV) M.Jenkins SCH.
              and Rock Outcrop 500'E. (Ref.(20)).

In an earlier section of this paper it was shown
that the characteristics of strong ground motions are
dependent on various parts of the source-to-site chain.
Since the generation, propagation and order of magnitude of microtremor motions are not similar to the comparable features of a strong motion earthquake, there
is no reason to believe that a direct correlation between the two processes should exist. Furthermore, the
known non-linear stress-strain relations of soil properties (damping, modulus etc.) precludes the direct
extrapolation of microtremor results, which relate to
conditions in the infinitesimal (linear) strain range,
to predict site characteristics during major earthquakes when non-linear behaviour is admitted. It is
for reasons such as these that the microtremor method
is open to much valid criticism, and interferences
drawn from such studies should be treated with caution.

Some of the difficulties noted above are under-
lined in the results of a recent study by Udwadia and
Trifunac.[23] These authors compared the spectra of a
number of different strong motion earthquakes recorded
at a particular site with the spectra obtained from
microtremor measurements in the same site region. The
investigation indicated that microtremor processes are
of a different character from earthquake processes. The
former show a broader spectrum with major energy content at relatively higher frequencies; they seem to
give details of the input creating the motions rather
than the transfer function of the site. Furthermore,
as may be seen in Figure 26, which compares microtremor
spectra derived from measurements taken at 24 hour intervals, the motions are nonstationary and therefore
the spectral peaks are not consistently repeatable. Accordingly, it does not seem generally meaningful at
this stage to attach special significance to the
spectral pattern of microtremors, in terms of their representation of site periodicity or response under
strong ground shaking. Further studies are necessary
and correlations with earthquake motions, if they exist,
must be clearly identifiable before the process can be
considered as an acceptable indicator of dynamic site
characteristics.

ACKNOWLEDGEMENTS

No claims to originality are expressed by the
author of this material; he has drawn heavily on the
works of others in its preparation. Numerous published

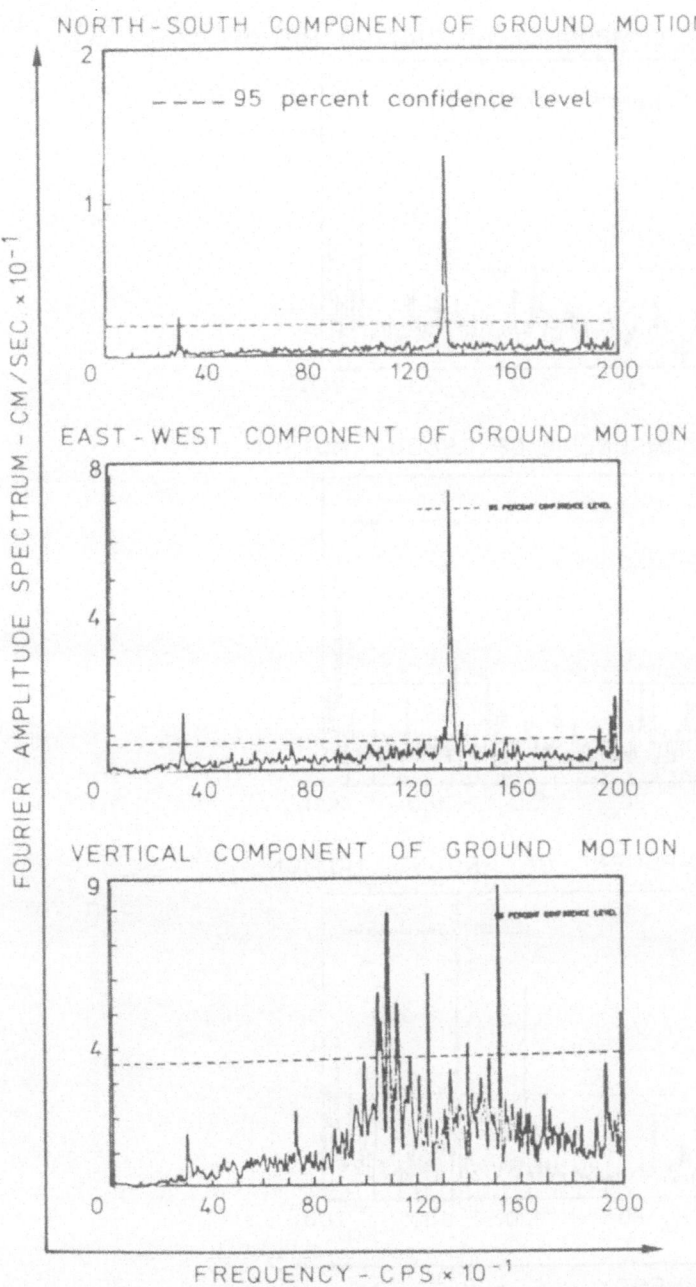

Figure 26a: Fourier analysis of microtremor acceler-
             ation measurements. El Centro Cityhall,
             August 4., 1970, 10 PM. (Ref.23).

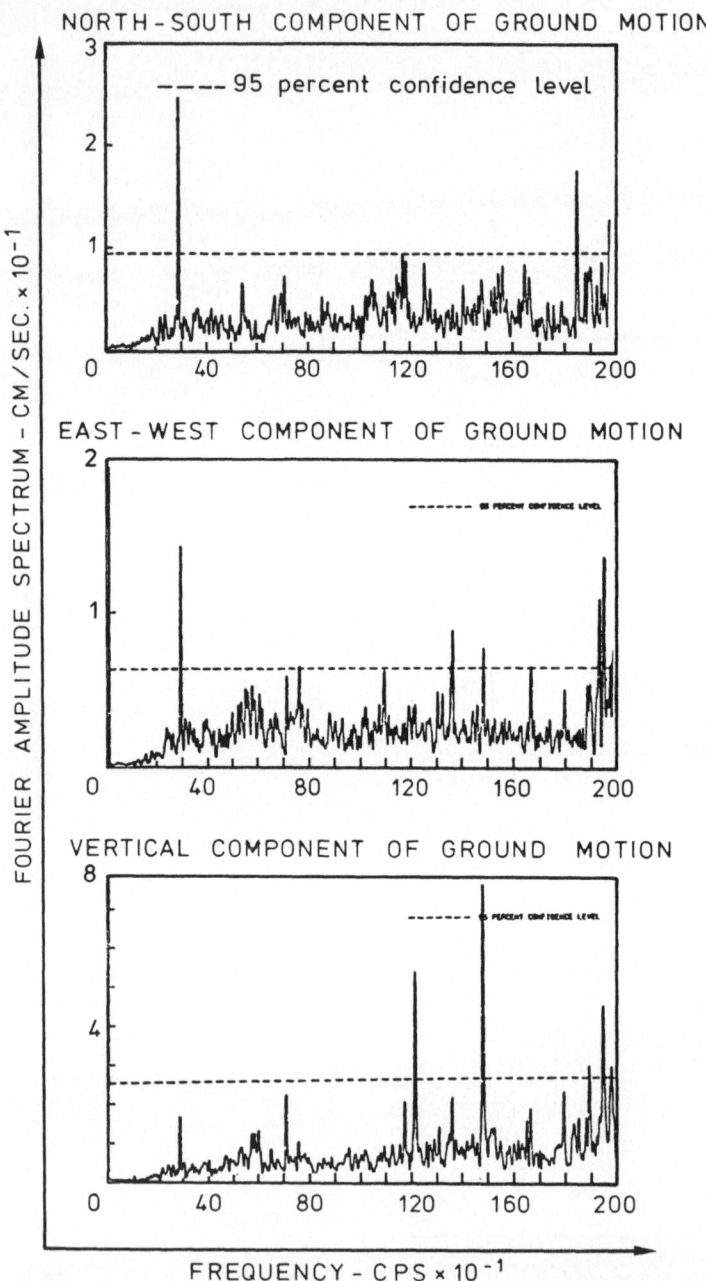

Figure 26b: Fourier analysis of microtremor acceleration measurements 24 hours later. El Centro Cityhall, August 25., 1970, 10 PM. (Ref.23).

papers by Hudson and Housner have been particularly helpful and this fact is gladly acknowledged.

REFERENCES

1. Halverson, H.T., Modern trends in strong movement (Strong-Motion) instrumentation. Dynamic Waves in Engineering, J.Wiley, 341, 1971.

2. NOAA, Rm. 7076, 390 Main St., San Francisco, California.

3. ERI, University of Tokyo, Japan.

4. Hudson, D.E. and Cloud, W.K., An analysis of seismoscope data from the Parkfield earthquake of June 27, 1966. Bull. Seism. Soc.America, 57, 1143, 1967.

5. Hudson, D.E., Ground Motion Measurement, Earthquake Engineering, Prentice Hall, 1970.

6. Halverson, H.T., The strong-motion accelerograph, 3rd World Conf. Earthq.Eng. 1, III-75, 1965.

7. Whitman, R.V., Basic concepts and important problems, Seismic Design for Nuclear Power Plants - MIT Press 1970.

8. Perez, V., Velocity response envelope spectrum as a function of time, for the Pacoima dam, San Fernando earthquake, Feb. 9, 1971. Bull. Seism. Soc. America, 63, 299, 1973.

9. Housner, G.W., Behaviour of structures during earthquakes, Proc. ASCE, Journal of the Eng. Mech. Div., 85, 1091, 1959.

10. Housner, G.W. and Jennings, P.C., Problems in seismic zoning, Paper No.203.5th World Conf. Earthq. Eng., 1973.

11. Housner, G.W., The role of earthquake ground motions in earthquake engineering, Invited Paper, 5th World Conf. Earthq. Eng., 1973.

12. Cloud, W.K. and Perez, V., Strong motion records and acceleration, 4th World Conf.Earthq.Eng., Chile 1969.

124

13. Hudson, D.E., Strong motion seismology, Proc. of Int. Conf. on Microzonation, 1, 29, Seattle,1972.

14. Housner, G.W., Strong Ground Motion, Earthquake Engineering, Prentice Hall,1970.

15. Seed, H.B., Idriss, I.M. and Kiefer, F.W., Characteristics of rock motions during earthquakes, Dept. No. EERC 68-5. College of Eng. U.of California, Berkeley, 1968.

16. Hudson, D.E. and Udwadia, F.E., Local distribution of strong earthquake ground motions, Paper No.78, 5th World Conf. Earthq. Eng., Rome 1973.

17. Jennings, P.C., Housner, G.W. and Tsai, N.C., Simulated earthquake motions for design purposes, 4th World Conf. Earthq. Eng., Chile 1969.

18. Newmark, N.M. and Rosenblueth, E., Fundamentals of Earthquake Engineering, Prentice Hall, 1971.

19. Schnabel, P., Seed, H.B. and Lysmer, J., Modification of seismograph records for effects of local conditions, EERC 71-8 U. of California, Berkeley.

20. Cherry, S. and Salt, P.E., A preliminary investigation of microtremor spectra, Proc. 1st Canadian Conf. on Earthquake Eng. Vancouver, 1971.

21. Kanai, K., On the spectrum of strong motion earthquakes, Primeras Journadas Argentinas De Ingenieria Antisismica, XXIV-1, 1962.

22. Alcock, E.D., Grand Valley Colorado: A microzonation case history, Proc. Int.Conf. on Microzonation, I, Seattle, 1972.

23. Udwadia, F.E. and Trifunac, M.D., Studies of strong earthquake motions and microtremor processes, Proc. Int.Conf. on Microzonation, I, Seattle, 1972.

ESTIMATING UNDERGROUND MOTIONS FROM SURFACE ACCELEROGRAMS

by S.Cherry
Professor of Civil Engineering
University of British Columbia
Vancouver, Canada

SUMMARY

A qualitative description is provided for estimating underground seismic motions from recorded surface accelerograms. The Fourier transform technique and wave propagation solutions are outlined and the inhomogeneous and viscoelastic nature of the soil are contemplated in discussing the dynamic response of a layered system. The results of the governing theory are shown through application to a number of strong motion records and site conditions. The material presented is based on information taken from various sources in the literature and is intended as a simple supplement to such work.

INTRODUCTION

The response of a horizontal soil profile to vertically incident, transient shear waves may be calculated using both lumped mass solutions and wave propagation techniques. The latter method offers a closed-form solution which can be conveniently and efficiently obtained by employing a Fast Fourier Transform algorithm.

The multiple reflection theory of waves in a layered system is qualitatively outlined below, in a very general sense, to compliment the heavy theoretical presentation of the accompanying references 1,2,3,4,5 on

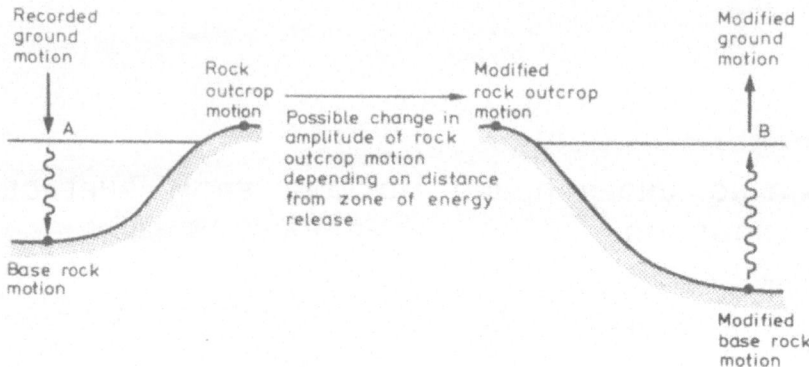

Figure 1: Schematic representation of a procedure for
computing effects of local soil conditions
on ground motions (Ref.2).

the inference of underground motions from surface accelerograms. The motions at any depth below the ground
surface, down to and including bedrock, can be estimated
by this approach. As a logical extension of this process, the motions of adjacent rock outcrops and adjacent layered sites exhibiting different soil conditions
can similarly be estimated in the manner suggested by
Schnabel, Seed and Lysmer [2] and illustrated graphically
in Figure 1. It must be emphasized that the technique
pre-supposes a very simplified wave propagation concept
which is open to valid criticism.

BASIC PRINCIPLES

Fourier Transform Technique

    The wave propagation solution for the transient
response of a system involves the following of basic
steps:
    (1) computation of the transfer function of the
system or ratio of steady state amplitude of output to
that of input as a function of the frequency.
    (2) transformation of the input function from the
time to the frequency domain by Fourier transform techniques.
    (3) multiplication of the Fourier transform of the

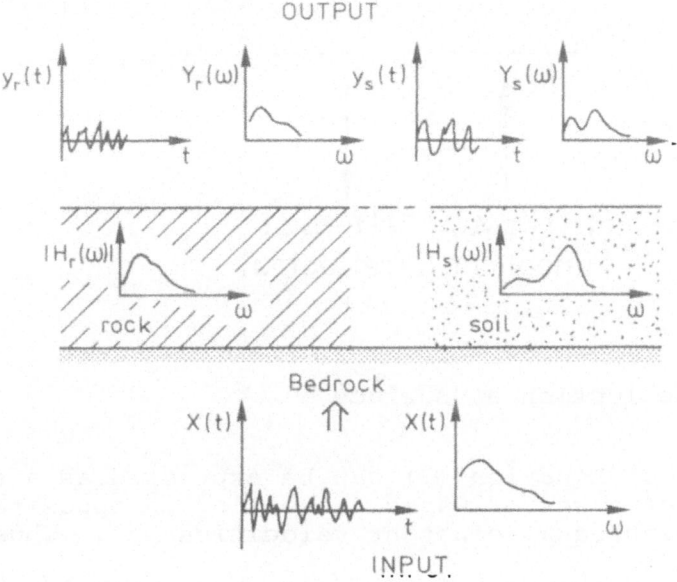

OUTPUT

INPUT

Figure 2:   Modification of baserock motion

input motion by the transfer function of the system to
yield the Fourier transform of the output motion.
    (4) inversion of the results of step 3 to yield
the time history of the output.

    Figure 2 offers a diagrammatic illustration of
this process in terms of a transient input motion x(t)
at bedrock. H(ω) represents the transfer function and
y(t) the output function.

Shear Wave Propagation

    For vertically incident shear waves in an elastic
homogeneous medium, the transverse deformations u as
a function of time t and depth z , Figure 3, are
governed by the one-dimansional wave equation

$$\frac{\partial^2 u}{\partial t^2} - c^2 \frac{\partial^2 u}{\partial z^2} = 0 \qquad (1)$$

where $c^2 = \mu/\rho$ = shear wave velocity

and μ and ρ = material shear modulus and density
respectively.

128

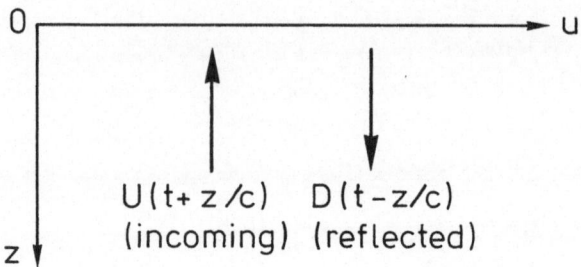

Figure 3:   Reflection at surface

The solution of Equation (1) can be expressed as a superposition of a set of waves travelling in opposite
directions with equal constant velocities c . Thus,

$$u(z,t) = U(t + z/c) + D(t-z/c) \tag{2}$$

where the functions U and D define waves advancing in
the negative z-direction (travelling up) and the positive z-direction (travelling down) respectively.

Surface Reflected Waves

     On reaching a free surface the ascending wave is
reflected downward and the actual displacement of the
ground at any position z is given by Equation (2). U
and D are related as may be seen by applying the boundary condition specifying zero shear stress (or strain)
at the free surface.

     Thus, from Equation (2)

$$\partial u/\partial z = \text{shear strain} = \frac{1}{c} U'(t+z/c) - \frac{1}{c} D'(t-z/c)$$

and from $\partial u/\partial z = 0$ at $z = 0$ it follows that

$$D'(t) = U'(t)$$

or

$$D(t) = U(t) \tag{3}$$

The primes denote differentiation with respect to z .

Equation (3) is a statement of the fact that the wave of the incident and reflected waves are of the same shape and amplitude and hence,

$$u(z,t) = U(t-z/c) + U(t+ z/c) \qquad (4)$$

At the surface (z = 0)

$$U(0,t) = u_o(t) = 2U(t) \qquad (5)$$

or

$$U(t) = u_o(t)/2$$

indicating that the free surface amplitude is twice that of the incident wave.

Wave Reflection and Transmission at a Layer Boundary

When impinging on a boundary between two differring strata, Figure 4, a portion, U_1, of the incident wave U_2 is transmitted upward and a portion is reflected downward, D_2 . The subscripts refer to the designated layers and the layer properties are noted on the diagram. The motion in each layer follows directly from Equation (4) and can be written as:

$$\left. \begin{aligned} u_1(t,z) &= U_1(t+ z/c_1) + D_1(t- z/c_1) \\[2ex] u_2(t,z) &= U_2(t+ z/c_2) + D_2(t- z/c_2) \end{aligned} \right\} \qquad (6)$$

To maintain the required continuity along the common boundary between the layers the following conditions must hold true

$$u_1(t,0) = u_2(t,0) \ldots \text{equal displacement}$$

and

$$\mu_1 \left(\frac{\partial u_1}{\partial z}\right)_{z=0} = \mu_2 \left(\frac{\partial u_z}{\partial z}\right)_{z=0} \ldots \text{equal shear stress}$$

When these boundary conditions are imposed it is possible to express

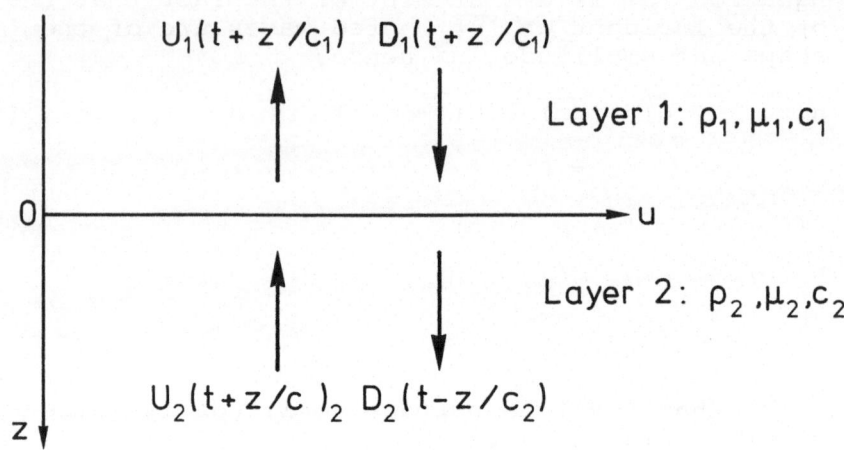

Figure 4: Transmission and reflection at interface

$$U_1 \quad \text{in terms of} \quad U_2 \quad \text{and} \quad D_1$$

and

$$D_2 \quad \text{in terms of} \quad U_2 \quad \text{and} \quad D_1$$

by means of the layer constants c and ρ . The actual solutions are not completed here.

Single Layer Over Bedrock Excited by Steady-State Input

Figure 5 represents a uniform soil layer of thickness H underlain by an elastic half-space of rock. A steady-state shear wave with incidence wave amplitude a_1 and frequency ω travelling up through the rock can be written in the form

$$u = a_2 e^{i\omega(t+ z_2/c_2)} = a_2 e^{i(\omega t+k_2 z_2)} \tag{7}$$

where

$$k_2{}^2 = \frac{\omega^2}{c_2{}^2} = \frac{\omega^2 \rho}{\mu} \tag{8}$$

is the propagation constant in the rock. The displacement of a point in the rock (material 2) and in the soil layer (material 1) can be expressed as:

$$u_1 = U_1 + D_1 = a_1 e^{i(\omega t + k_1 z_1)} + b_1 e^{i(\omega t - k_1 z_1)}$$

and

$$u_2 = U_2 + D_2 = a_2 e^{i(\omega t + k_2 z_2)} + b_2 e^{i(\omega t - k_2 z_2)}$$

$$\left.\vphantom{\begin{array}{c}1\\1\\1\end{array}}\right\} \quad (9)$$

At the free surface $(z_1 = 0)$, the boundary condition statement of zero shear is

$$(\partial u_1 / \partial z_1)_{z_1 = 0} = 0$$

and when the first of Equations (9) is substituted this yields

$$a_1 = b_1 \qquad (10)$$

leading to

$$u_1(z_1, t) = a_1 e^{i(\omega t + k_1 z_1)} + a_1 e^{i(\omega t - k_1 z_1)}$$

This indicates that the incident and reflected waves have the same amplitude. The displacement at the surface $(z_1 = 0)$ becomes

$$u_0(t) = 2a_1 e^{i\omega t} \qquad (11)$$

which is double the amplitude of the incident motion.

From the two additional boundary conditions expressing the continuity of displacement and shear at the contact surface between the soil and rock we can write

$$(u_1)_{z_1 = H} = (u_2)_{z_2 = 0} \qquad (12a)$$

and

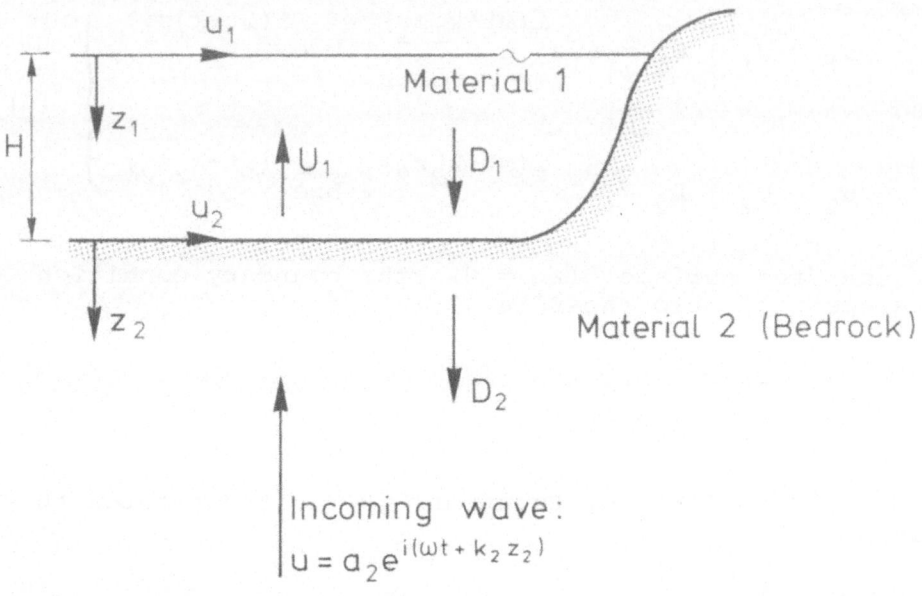

Figure 5:   Single layer over bedrock

$$\mu_1 (\partial u_1 / \partial z_1)_{z_1 = H} = \mu_2 (\partial u_2 / \partial z_2)_{z_2 = 0} \qquad (12b)$$

On substituting Equations (9) into Equations (12) and recalling Equation (10) we obtain

$$\left. \begin{aligned} a_2 &= a \; a_1 \\ b_2 &= b \; a_1 \end{aligned} \right\} \qquad (13)$$

where a and b are coefficients involving H, ρ, μ and ω .

We can now express the ratio between the displacement (acceleration) amplitude at the free surface of the soil and the underlying rock surface as (from Equations 9, 10 and 13):

$$\frac{(u_1)_{z_1=0}}{(u_2)_{z_2=0}} = \frac{2a_1}{a\,a_1+b\,a_1} = \frac{2}{a+b} \tag{14}$$

The corresponding ratio between the soil and rock out-crop surfaces is

$$\frac{(u_1)_{z_1=0}}{(u_2)_{z_2=H}} = \frac{2a_1}{2a_2} = \frac{a_1}{a_2} = \frac{1}{a} \tag{15}$$

Multi-layered Viscoelastic Site

The preceding sections assumed an elastic homogeneous material behaviour. In fact, the material has certain damping properties. The governing differential equation for the response of the j'th layer of a multi-layered site, Figure 6, in which the energy dissipative properties are accounted for by assuming viscoelastic behaviour of the material is:

$$\rho_j \frac{\partial^2 u_j}{\partial t^2} - \mu_j \frac{\partial^2 u_j}{\partial z^2} - \nu_j \frac{\partial^3 u_j}{\partial z^2 \partial t} = 0 \tag{16}$$

where ν_j represents the coefficient of viscosity of the material. Damping is represented by the last term in this equation. Following the identical procedure outlined above, the solution for the various layers corresponding to a steady-state incoming wave expressed as:

$$u = a_{n+1} e^{i(\omega t + k_{n+1} z_{n+1})}$$

is

$$u_1 = a_1 e^{i(\omega t + k_1 z_1)} + b_1 e^{i(\omega t - k_1 z_1)}$$

$$\vdots$$

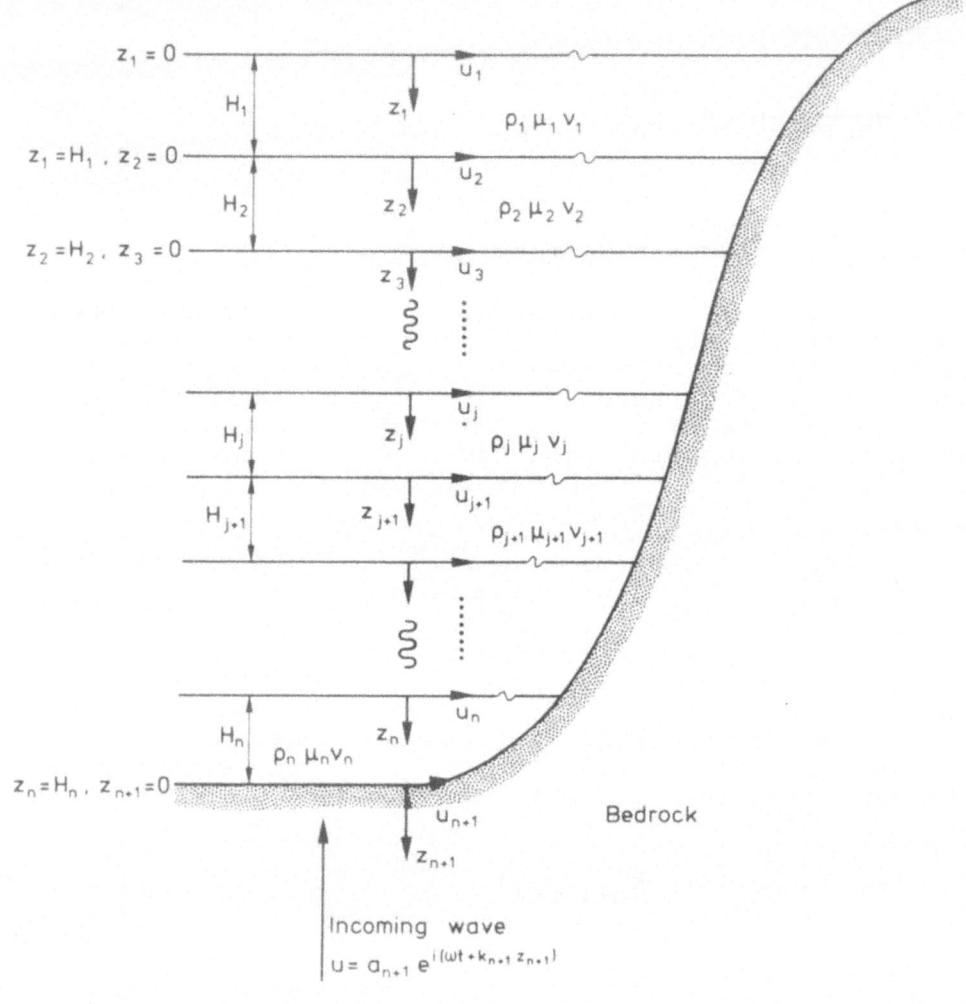

Figure 6: Viscoelastic multi-layer site over bedrock.

$$u_j = a_j e^{i(\omega t + k_j z_j)} + b_j e^{i(\omega t - k_j z_j)} \left.\begin{array}{c} \\ \\ \\ \\ \\ \end{array}\right\} \quad (17)$$

$$u_{j+1} = a_{j+1} e^{i(\omega t + k_{j+1} z_{j+1})} + b_{j+1} e^{i(\omega t - k_{j+1} z_{j+1})}$$

where the propagation constant k_j for the j'th layer is:

$$k_j^2 = \frac{\rho_j \omega^2}{\mu_j + i\omega\nu_j} \quad (18)$$

and

$$c_j^2 = \frac{\omega^2}{k_j^2} = \frac{\mu_j + i\omega\nu_j}{\rho_j} \quad (19)$$

is the complex shear wave velocity in the j'th layer.

On the basis of experiments $\omega\nu$ and μ are essentially independent of frequency for many soils so that c_j^2 can be re-written as:

$$c_j^2 = \mu_j(1 + 2i\beta_j)/\rho_j \quad (20)$$

where β_j is the fraction of critical damping for the j'th layer. The terms associated with each layer in Equations (13) represent the upward travelling transmitted wave and the downward travelling reflected wave.

Considering the top layer, the condition of zero shear at the free surface requires that

$$(\mu_1 + \nu_1 \frac{\partial}{\partial t}) \frac{\partial u_1}{\partial z_1} = 0 \qquad \text{at} \quad z_1 = 0$$

or

$$\mu_1 \frac{\partial u_1}{\partial z_1} = 0 \quad (21)$$

Setting the first of Equations (17) into Equation (21) leads to:

$$a_1 = b_1 \tag{22}$$

Hence,

$$u_1 = a_1 e^{i(\omega t + k_1 z_1)} + a_1 e^{i(\omega t - k_1 z_1)} \tag{23}$$

and the free surface motion is

$$u_o(t) = 2a_1 e^{i\omega t} \tag{24}$$

The surface motion amplitude is twice the incident wave amplitude.

Furthermore, at each interface the displacements and shears in the adjacent layers must be equal. This generates 2n boundary conditions at

$$z_j = H_j \quad \text{and} \quad z_{j+1} = o$$

defined by

$$u_j = u_{j+1}$$

and

$$\left(\mu_j + \nu_j \frac{\partial}{\partial t}\right) \frac{\partial u_j}{\partial z_j} = \left(\mu_{j+1} + \nu_{j+1} \frac{\partial}{\partial t}\right) \frac{\partial u_{j+1}}{\partial z_{j+1}} \tag{25}$$

$$\text{for} \quad j = 1 \dots n$$

Upon setting Equations (17) into Equations (25) a general recursion formula is obtained for a_j and b_j. From this we obtain

$$a_1 = b_1$$

and

a_2 and b_2 can be expressed in terms of a_1

a_3 and b_3 can be expressed in terms of a_2 and b_2

and hence in terms of a_1

$\vdots$

a_{n+1} and b_{n+1} can be expressed in terms of a_n and b_n

and hence in terms of a_1

or

$$a_{n+1} = a\ a_1$$
$$b_{n+1} = b\ a_1$$

(26)

where a and b are complicated coefficients involving ω and the ρ, μ, H and ν values of the layers.

From Equations (17), (24) and (26) the amplification A can be defined as

ratio of $\dfrac{\text{displacement (acceleration) at free surface}}{\text{displacement (acceleration) at underlying bedrock surface}}$

Thus

$$A = \frac{u_o}{u_{n+1}(0)} = \frac{2a_1}{a_{n+1}+b_{n+1}} = \frac{2a_1}{a\ a_1+b\ a_1} = \frac{2}{a+b}$$ (27)

Alternatively, the amplification, when described in terms of ratio of soil to rock outcrop surface amplitudes is

$$A = \frac{u_o}{u_{n+1}(0)} = \frac{2a_1}{2\ a_{n+1}} = \frac{2a_1}{2\ a\ a_1} = \frac{1}{a}$$ (28)

which represents the ratio of surface motion with layering to that without layering.

While the general expression for A becomes very complicated, a systematic numerical calculation and progression through the system from layer to layer is readily solved by digital computation.

Non-Homogeneous Elastic Case

The inhomogeneous condition of the soil can be accounted for by permitting the shear modulus to vary with depth. The basic equation governing motion of a non-homogeneous elastic material is:

$$\rho \frac{\partial^2 u}{\partial t^2} - \frac{\partial \mu(z)}{\partial z} \frac{\partial u}{\partial z} - \mu(u) \frac{\partial^2 u}{\partial z^2} = 0$$

where $\mu(z)$ is the variable shear modulus which can be assumed to vary in some way with the depth of soil. Thus, assuming a variation as the reciprocal of the n'th power of the depth we can write:

$$\mu(z) = \mu_0 (z + d)^{1/n}$$

where μ_0 is a proportionality constant and d is a constant describing the shear modulus at the free surface. The general solution for the resulting Bessel's differential equation is complicated and expressed as a transform [1] which involves the Fourier transform of the recorded accelerogram. The underground response is then evaluated by normal inverse transform procedures.

Transient Input and Non-Linear Soil Properties

The discussions outlined above are mainly concerned with steady-state inputs and responses. Transient motions exhibiting earthquake characteristics can be handled by the Fourier transform technique involving the four basic steps previously noted. Because soil properties are strain dependent, average curves showing the variation of the moduli and damping of the materials with strain, Figure 7, which was taken from reference 2, should be used in the dynamic analysis. In this connection, the appropriate parametric values which govern the true solution can be determined by iterating until a compatibility between the derived and assumed strains in the soil profile are obtained.

EXAMPLES OF APPLICATION OF ANALYSIS

Figure 8, reproduced from reference 2, is used to summarize, through examples, some of the results of the theory outlined above. It also indicates how local soil conditions can modify accelerograms. The recorded acceleration-time history due to the 1957 San Francisco earthquake was used to compute the underlying bedrock motion and the rock outcrop motion at the SP site. Attenuation curves were employed to estimate the rock outcrop motions near site SB (which was about 1½ miles

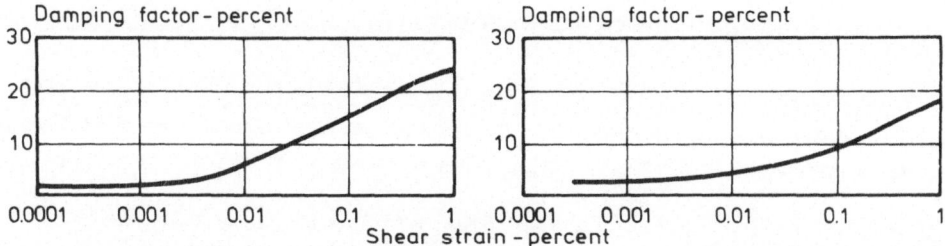

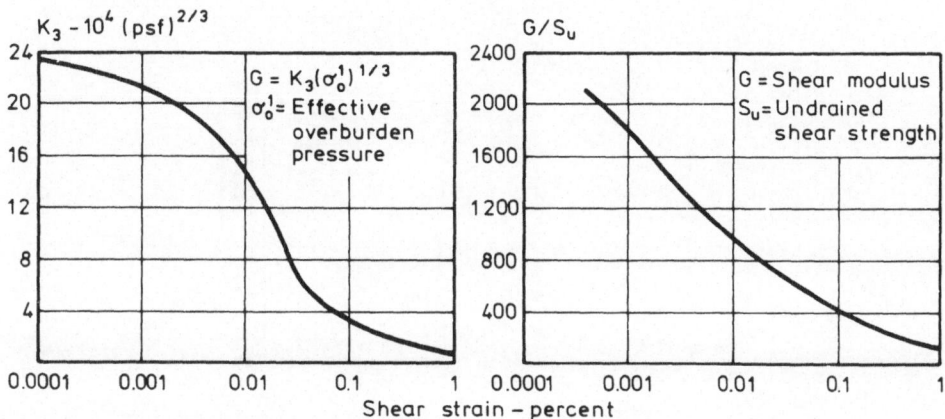

Figure 7a: Shear moduli and damping characteristics for sands at relative density of about 75 percent. (Ref.2).

Figure 7b: Shear moduli and damping characteristics for saturated clays. (Ref.2).

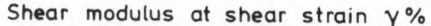

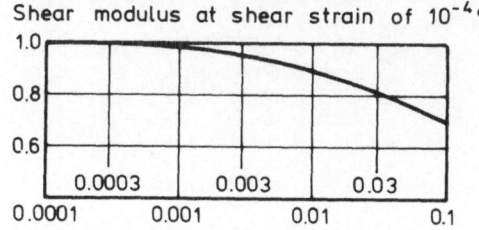

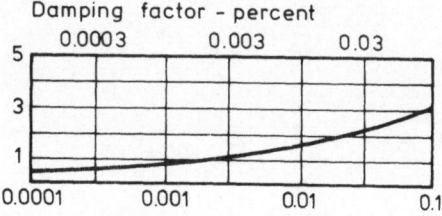

Figure 7c: Shear moduli and damping characteristics for rock. (Ref.2).

140

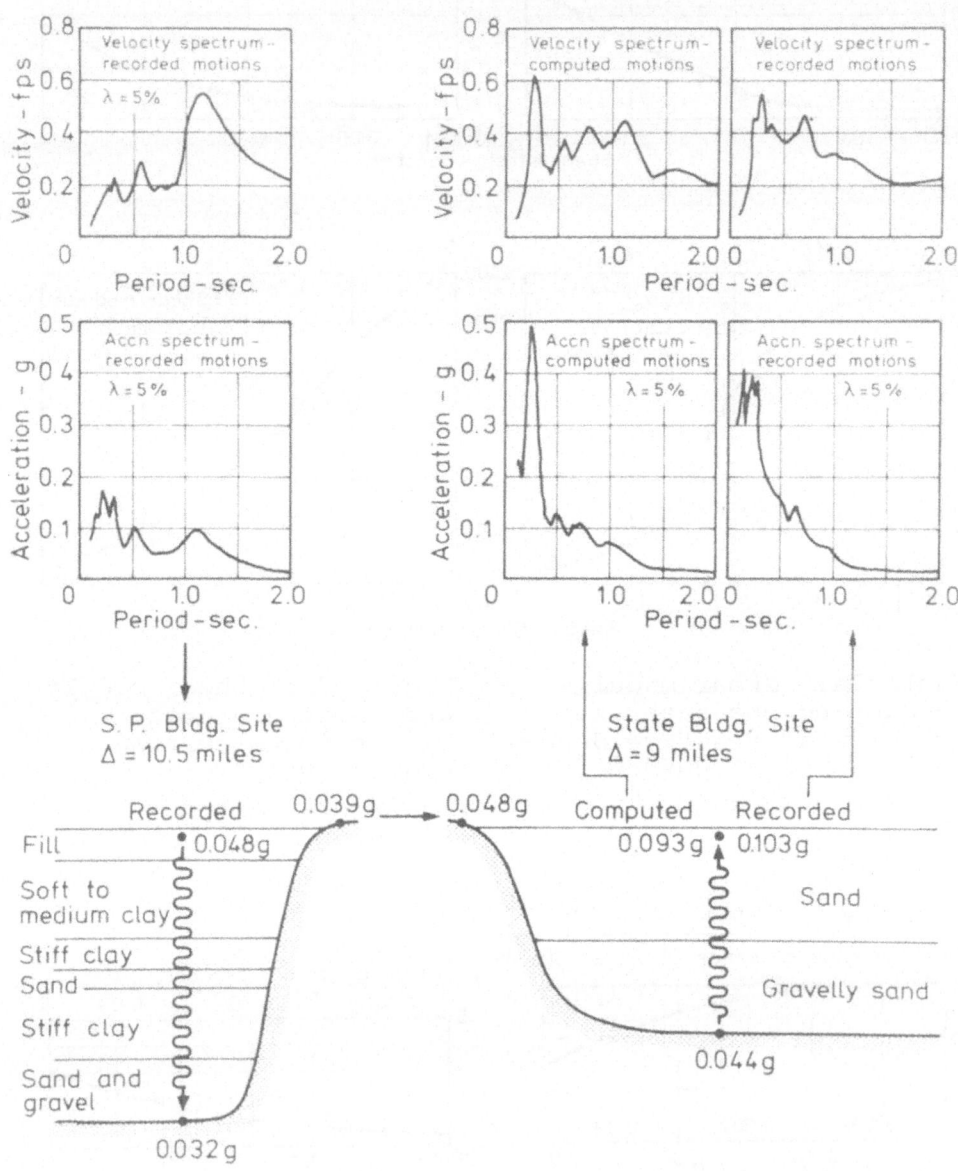

Figure 8: Response of state building site computed
from motions recorded at Southern Pacific
building site. (Ref.2).

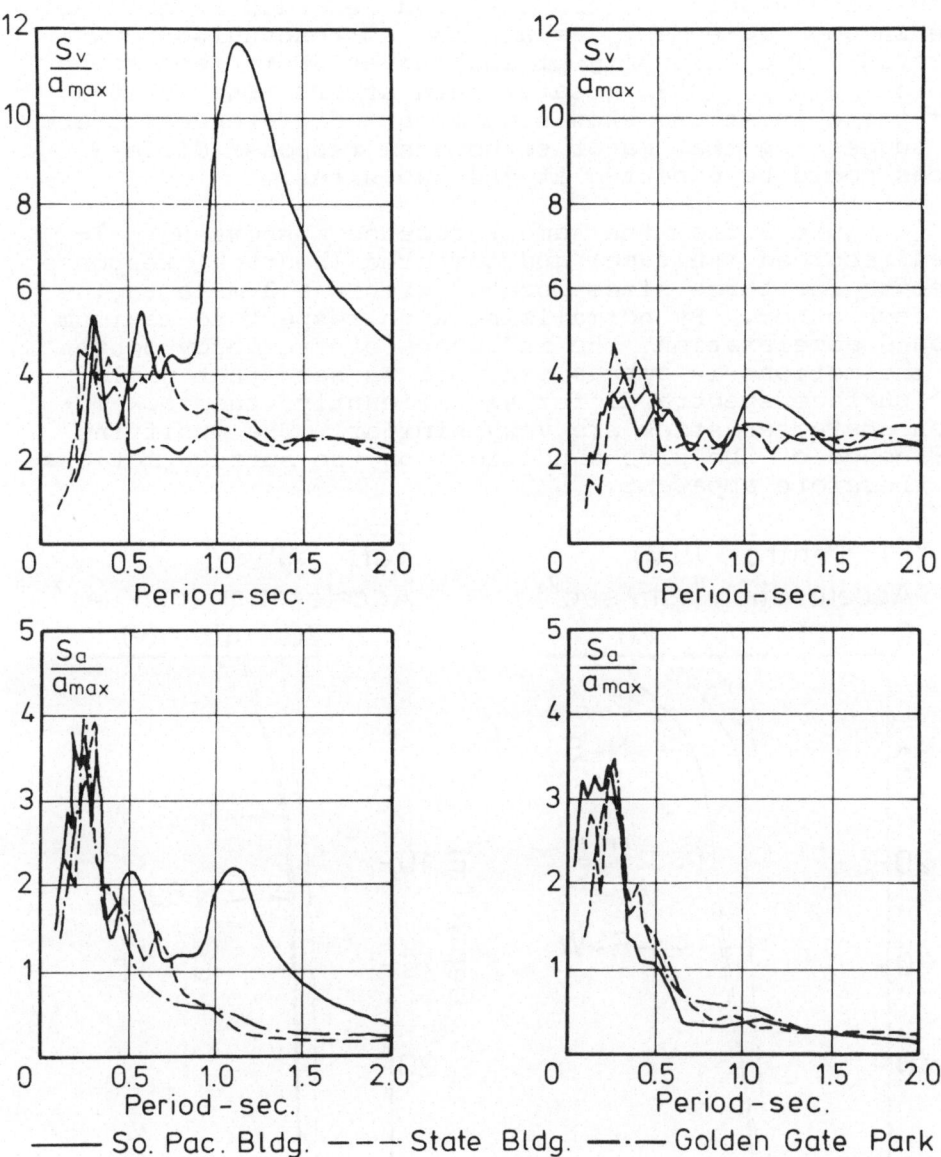

Figure 9a: Normalized response spectra for recorded motions (Ref.2).

Figure 9b: Normalized response spectra for computed rock motions. (Ref.2).

closer to the earthquake epicentre than the SP site),
from which the surface motions at that location were
then calculated. The computed and recorded SB motions
are in good agreement, as are the corresponding response
spectra. The intensity of shaking at SB is seen to be
approximately 100% greater than at SP. The spectra
for these locations show significant differences, thereby suggesting that major structural response differences could be expected at the two sites.

Figure 9 from the same reference [2] shows normalized recorded (surface) and computed (bedrock) response
spectra for three sites located within a 3 mile radius
of each other. By normalizing with respect to maximum
ground acceleration, the influence of frequency content
of the motions is emphasized. It is seen that while
the surface spectra differ significantly, the rock motion characteristics are very similar. The modifying
influence of the site conditions on the surface motions
is therefore apparent.

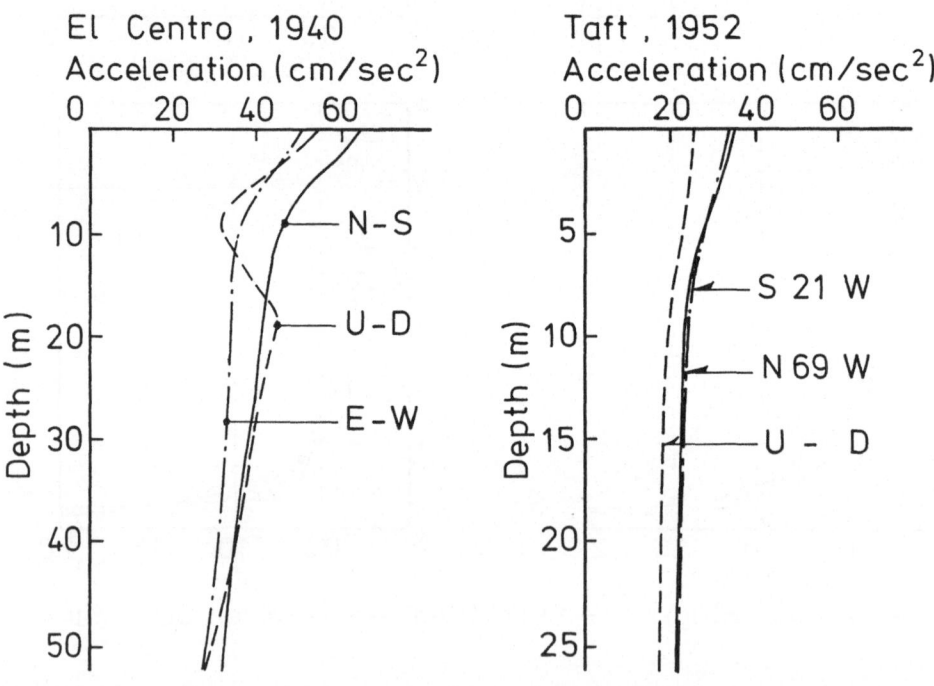

Figure 10:   R.M.S.Distribution of acceleration (ref.1)

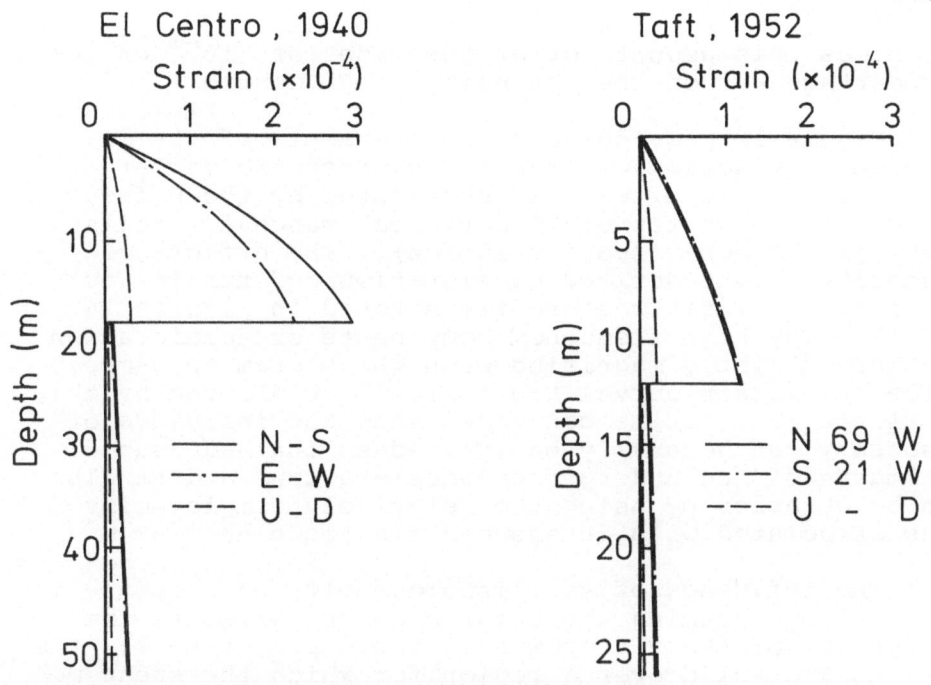

Figure 11: R.M.S.Distribution of strain (ref.1)

Figures 10 through 17 summarize the results of an investigation reported by Toki and Cherry [1] which was concerned with a method for estimating the underground motion induced during an earthquake through an analysis of the seismic record obtained at ground surface. In that study, theoretical solutions to this problem are provided for ground behaviour assumed as (1) elastic homogeneous; (2) viscoelastic homogeneous and;(3) non-homogeneous elastic. The theory is applied to an analysis of nine strong motion accelerograms recorded on the ground surface at five different sites. Some of the results are summarized below; the calculations were based on a two-layer model of the site conditions.

Typical plots of the r.m.s. distribution of acceleration and strain for the elastic homogeneous case are shown in Figures 10 and 11. The magnification of acceleration amplitudes is about 1.5 times and is seen to be concentrated in the near surface regions of the sites between the first interface (19 m and 12 m for El Centro and Taft respectively) and the ground surface. The r.m.s. of strain is of the order of 10^{-4} ; the maximum value of shear strain has been shown to be about

144

3-5 times this amount, or of the order of 10^{-3} or less, for earthquakes of the intensity of El Centro.

Figure 12 provides a typical example of the influence of viscosity on the r.m.s. response variations. The viscoelastic effect is represented by the D-factor; the values shown represent a typical range for soils (D=0 is the non-viscous solution). The effects of viscosity on the derived acceleration and strain records at the interface are illustrated in Figures 14 and 15. The high frequency components of acceleration are very slightly magnified when the system is damped, while the strain curves are basically unaltered by this consideration. It is concluded that the influence of viscosity can generally be ignored so that adequate estimates of the underground accelerations and strains can be obtained by using the relatively simple solutions associated with an assumed elastic behaviour.

The influence of soil inhomogeneity on response is examined by assuming the shear modulus varies as the reciprocal of the n'th power of the depth of soil; the results are valid over a region for which the shear mo-

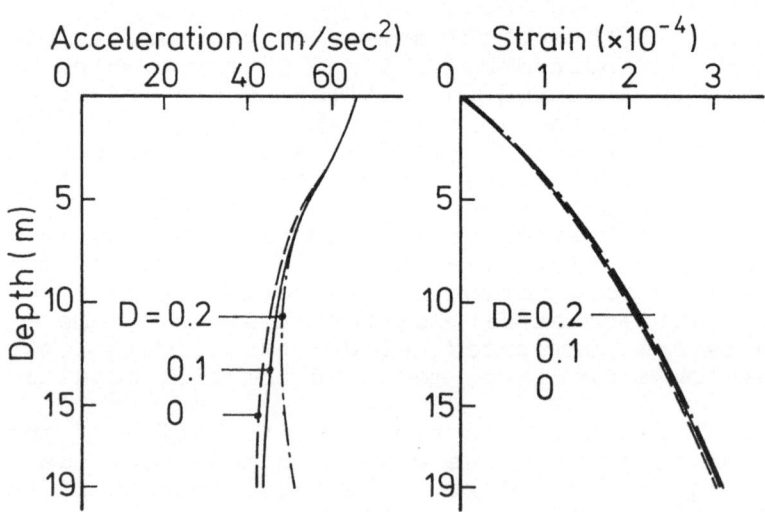

Figure 12: Effect of viscosity (EL Centro, 1940 N - S) (ref. 1)

dulus does not vary rapidly with depth. An example of the effect of inhomogeneity is contained in Figure 13. The r.m.s. distribution of acceleration is not materially influenced by the large differences in the wave velocities at the surface and the interface at 10 m depth (below which the velocity is assumed constant). However, there are notable differences in the r.m.s. strain variation for the homogeneous and the inhomogeneous cases; this quantity is strongly dependent on the wave velocity distribution and the velocity in the deeper soil, but not on the power of n . These general remarks apply also to the response variations in the time domain, as illustrated in Figures 16 and 17. It is concluded that the ground should be considered as a non-homogeneous medium when inferring the underground strains from the surface records, but that a simple, homogeneous model is adequate for providing reasonable estimates of the in-depth accelerations.

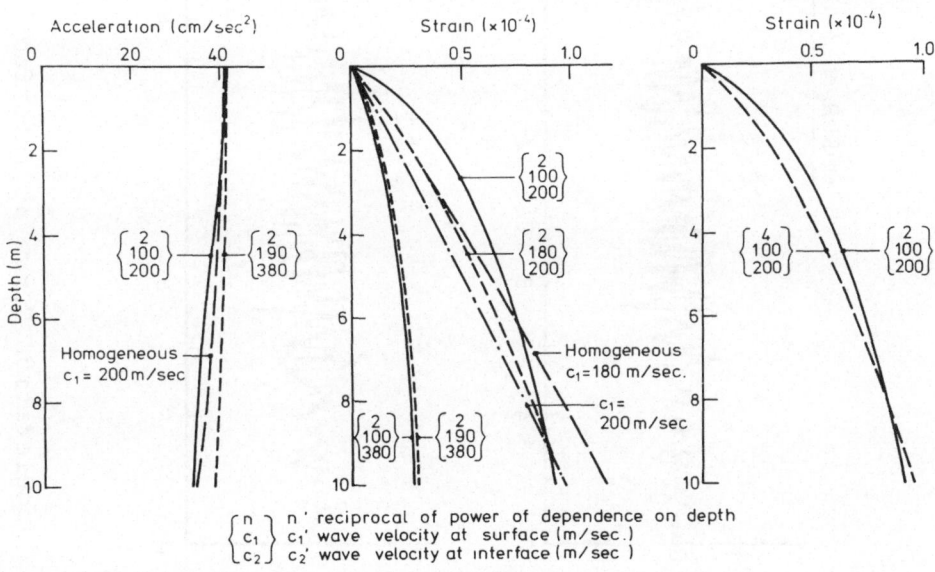

Figure 13:   Effect of inhomogeneity on R.M.S.distribution.(Hachinohe 1968 N-S). (ref.1).

146

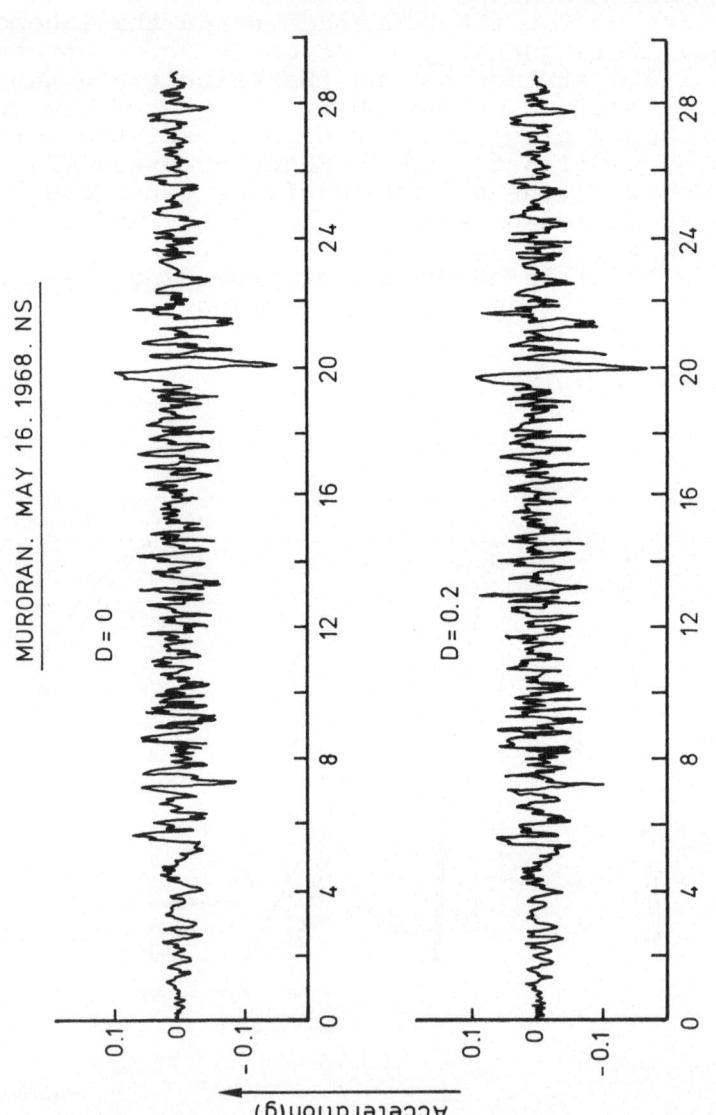

Figure 14: Acceleration Curve at Interface (ref.1)

147

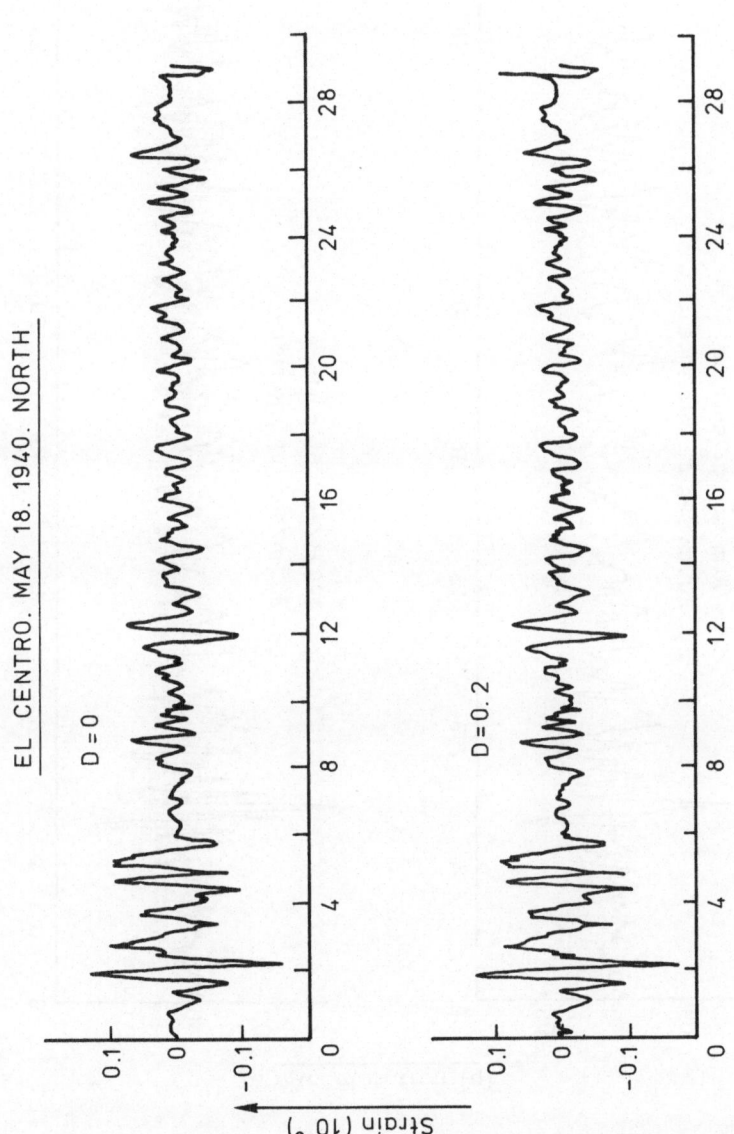

Figure 15: Strain curve at interface (ref.1)

148

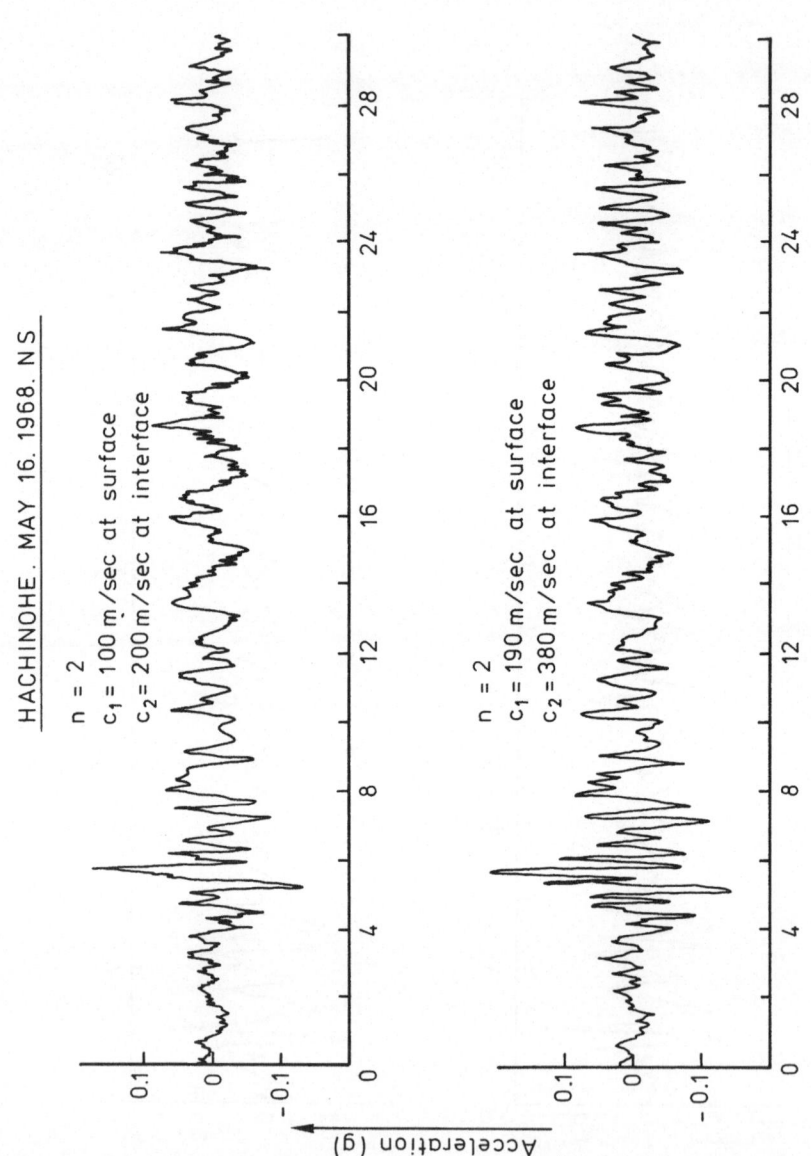

Figure 16: Acceleration curve at interface (ref.1)

149

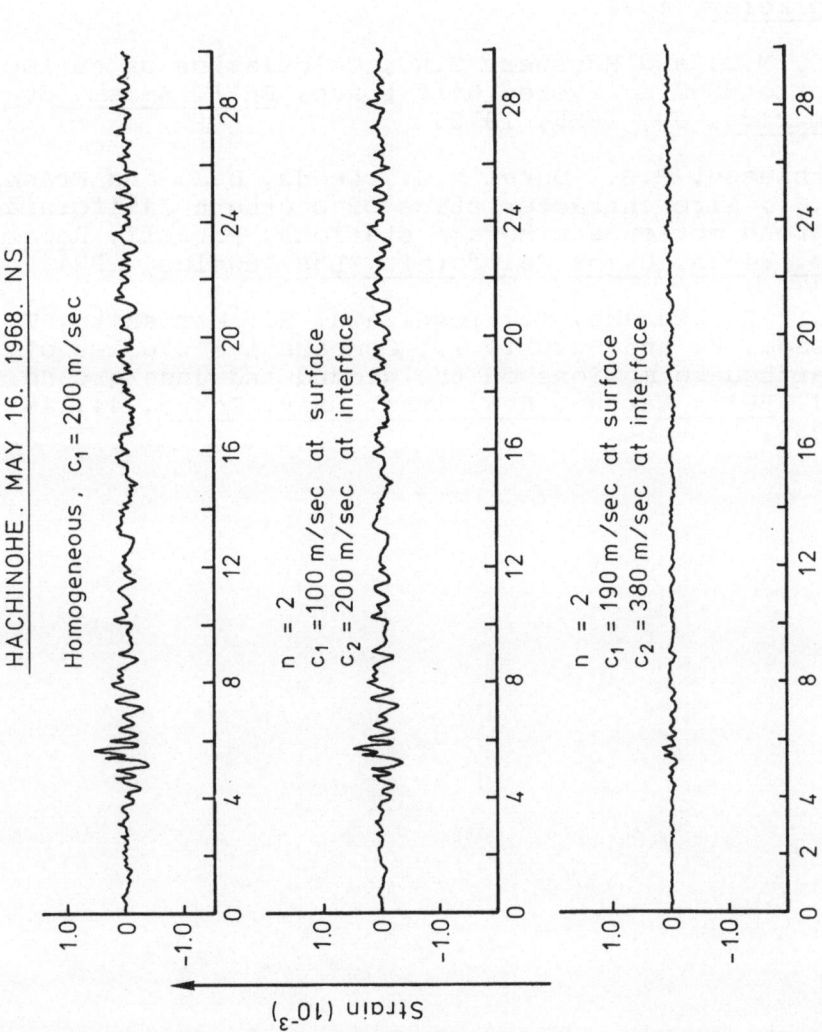

Figure 17: Strain curve at interface (ref.1)

150

REFERENCES

1 Toki, K. and Cherry, S., Inference of subsurface acceleration and strain from accelerograms recorded
   at ground surface, Proc. 4th European Conf. Earthq.
   Eng., London, 1972.

2 Schnabel, P., Seed, H.B. and Lysmer, J., Modification of seismograph records for effects of local
   soil conditions, EERC 71-8, U. of California,
   Berkeley, 1971

3 Tsai, N.C. and Housner, G.W., Calculation of surface
   motions of a layered half-space, Bull. Seism. Soc.
   America, 60, 1625, 1970.

4 Matthiesen, R.B., Duke, M.C., Leeds, D.J. and Fraser,
   J.C., Site characteristics of southern California
   strong motion earthquake stations, Part II, Report
   No. 64-15, U. of California, Los Angeles, 1964.

5 Kanai, K., Tanaka, T., Yoshizawa, S., Morishita, T.,
   Osada, K. and Suzuki, T., Comparative studies of
   earthquake motions on the ground and underground
   II, Bull. Earthq. Res. Inst. Univ. Tokyo, 44, 187,
   1966.

DESIGN INPUT FOR SEISMIC ANALYSIS

by S.Cherry

Professor of Civil Engineering,
University of British Columbia,
Vancouver, Canada.

SYNOPSIS

Methods for determining the earthquake exposure at
a site are reviewed and the selection of appropriate
aseismic design criteria to meet this threat are outlined. The construction of smooth design spectra,
established on the basis of average and envelope response spectrum values, are described. Procedures for
extrapolating or modifying existing accelerograms to
satisfy the specified characteristics of a design ac-
cellerationtime history record for a particular site
are noted. The material presented is based on information taken from various sources in the literature and
is intended as a summary of procedures for selecting
design inputs.

INTRODUCTION

The main features influencing measured ground motions at a site resulting from a dislocation along some
distant fault have already been examined in an accompanying paper. [1] Such motions represent the input
function for a dynamic analysis, and once this is specified standard procedures of structural dynamics can be
used to establish the seismic response (forces and deformations) developed.

The analyst may choose to express the input in the
form of a design response spectrum or as a time-history
acceleration (accelerogram) record. The object of this
paper is to summarize methods by which these functions

may be established.

Many modern building codes actually describe the
design input in terms of a variable seismic coefficient,
which defines the equivalent lateral static loading to
be used in a lesser form of seismic analysis. The idea
is to attempt to rationalize the dynamic behaviour of
structures in a simplistic manner. Although adequate
for the design of a large class of structures, this
procedure does not represent a general, realistic aseismic design criteria. Code methods are normally quite
straightforward and will not be discussed here. Instead, attention will be directed to an understanding
of the hazard at a particular site and the selection of
appropriate aseismic design criteria to meet this
threat.

DETERMINATION OF SITE SEISMIC HAZARD

Since the design input controls or specifies the
forces and motions developed in a structure it must be
chosen to reflect the degree of exposure anticipated
for the site. The concept of a dual level of design or
threat represents the basic engineering philosophy in
establishing aseismic criteria. Thus, the structure
should be capable of experiencing, with little or no
damage, the maximum probable earthquake which could
reasonably be expected to occur during its lifetime. at
the same time, it should be safe from collapse under
the action of the maximum possible or credible earthquake which can realistically be judged as having only
a low probability of occurrence. These threats must
ultimately be converted into meaningful ground motion
parameters, and design spectra or accelerograms, in
which economic considerations involving study of the
initial cost of providing higher seismic resistance,
versus future costs of repairs and loss in the absence
of such provisions, should logically be incorporated.

The determination or choice of the maximum probable and possible events should be based on a seismic
site evaluation program in which data from geological,
seismological, geophysical and engineering sources are
examined. Geological environment studies delineate the
active (or inferred) fault patterns and types in the
region, the extent of fault movements and the distances
to causative faults from the site. Seismicity data in
the form of earthquake epicentre (Richter magnitude, M,
and distribution) maps indicate trends of probable

faults, provide the basis for certain statistical studies, and suggest duration of future strong ground
shaking. Isoseismal maps (Modified Mercalli, MM), particularly in relation to historical earthquakes, sometimes offer another or sole means of assessing the future intensity or degree of ground shaking and of judg-
ing the areal distribution of severe shaking. Soil
dynamic investigations, such as the determination of
soil profile densities, shear wave velocities and shear
moduli can indicate local soil features which may enhance or reduce the extent of bedrock motions. It is
often necessary to establish the strength and character
of the threat in the complete absence of strong motion
recordings, but when such data are available their correlation with MM readings can provide a meaningful
engineering index of the size or intensity of this
threat.

From the seismic epicentre data noted above it is
generally possible to express certain magnitude - frequency laws for earthquakes in the region by a variety
of statistical risk studies such as those of Gumbel,[2]
Milne and Davenport,[3] Cornell and Vanmarcke,[4] Dick[5]
and Housner.[6] Estimates of the frequency of occurrence
of earthquakes of differing magnitudes and the probable
maximum magnitude of an earthquake occurring in a given
locality over a given time interval are examples of the
type of information which can be generated by such
studies. Similar relationships may also be developed
from MM intensity data, as has been done by Algermissen.[7]

Once this type of information is made available
seismic intensity recommendations (M and/or MM), corresponding to the maximum probable and maximum possible
earthquake motions for which the structure at a site
should be designed, can be specified. Related information such as distance to causative fault, duration of
strong ground shaking and predominant frequency content
can also be specified as part of the complete seismic
input; in this connection, the pertinent relationships
summarized in an accompanying paper [1] are very useful.

The recommended M and MM intensity values can
be converted to various ground motion parameters, such
as peak ground acceleration [6] and peak ground velocity[8]
and the attenuation of these quantities, as a function
of distance from causative fault, can be estimated by
means of a number of published attenuation relationships. Seed, Idriss and Kiefer [9] have summarized some

of these results and their recommended curves [1] permit
estimates of peak acceleration amplitudes and predominant frequency content to be made for rock and firm
ground sites.

Although the peak ground acceleration is frequently used to specify the seismic input, it is recognized
that this quantity, by itself, does not adequately or
appropriately describe the seismic threat or ground motion - duration of shaking and frequency content also
play an important role in establishing the system response. However, the relative maximum ground velocity
and displacement are usually inferred as proportional
to the peak acceleration and, as shown below, this
quantity can therefore be used as an index for specifying the shape of a design response spectrum. This
essentially has the effect of determining the frequency
content of the specified motion over the entire frequency regime. This index can also be used to modify
an existing time-history acceleration record, by proportional scaling or normalizing of its ordinates, to
arrive at a design earthquake satisfying the acceleration intensity requirements of a particular site.
Either the design spectrum or the design earthquake can
then be used to estimate the dynamic response of a
structure.

DESIGN RESPONSE SPECTRUM

The design spectrum is a plot of peak response of
a single-degree-of-freedom system as a function of period and damping. Once the damping and allowable stresses are prescribed the earthquake forces for which the
structure is to be designed can be determined from the
spectrum.

Housner [6] has established idealized or smoothed
spectra based on the average spectra shape of a number
of strong motion measurements recorded on firm ground
and corresponding to different earthquake magnitudes
and causative fault distances. Figure 1 reproduces
these curves for a peak ground acceleration index of
20% g . These spectra can be converted to other specified inputs (peak ground acceleration), appropriate to
the design intensity proposed for a site, by a proport-
inonal multiplication of spectrum ordinates. Thus, for
a 0.1 g seismic threat the scaling factor for any response quantity in Figure 1 would be 0.1 g/0.2 g .

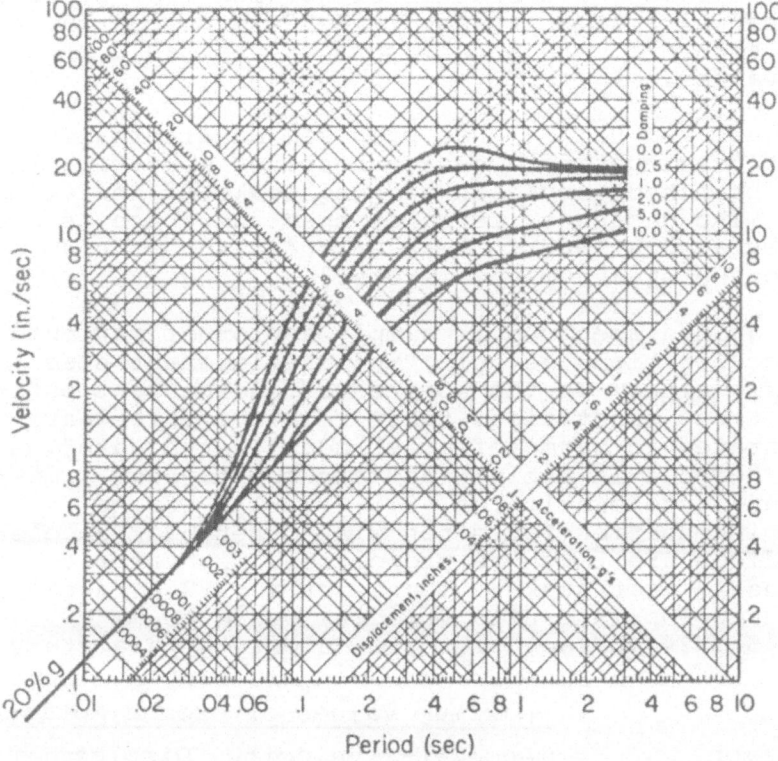

Figure 1: Combined Plot of Design Spectrum Giving S_a, S_r and S_d as a Function of Period and Damping; Scaled to 20% g Acceleration at Zero Period. (ref.6)

The use of smoothed spectrum to replace the rapidly fluctuating ordinates which characterize the spectrum of a real earthquake has definite advantages. It
overcomes the problem of making the design particularly
sensitive to the period estimates of a structure, which
can only be calculated with limited accuracy. In any
response calculation, the prescribed damping and allowable design stresses should relate to the manner in
which the design spectrum has been smoothed - whether
it represents a smooth curve of average response values
or an enveloping of the peaks of individual spectra. An
average spectrum does not account conservatively for
fluctuations about the mean which occur in a single
spectrum. When using average spectra the appropriate

allowable stresses are elastic or working load stresses, rather than ultimate stresses. The design spectra shown in Figure 1 were established on the basis of average values.

Newmark and Hall [10] has proposed an alternative method of constructing a design response spectrum. This employes branched curves enveloping peaks of individual spectra. The design criteria or spectrum shape is controlled by a relationship between the response and ground motion spectrum parameters.

The ground spectrum is simply a plot of maximum ground acceleration, velocity and displacement specified for a site. In the absence of independent estimates of these quantities, their relative values are usually stated in terms of a "standard earthquake", which is defined in Table 1 and assumed to have a maximum ground acceleration of 0.5 g . The ground spectrum intensities for sites with other peak acceleration indices are taken as proportional in value to this standard earthquake. Thus, for a 0.2 g acceleration the appropriate site velocity would be (0.2 g/0.5 g)24 in/sec. The ground motion parameters

Condition	Maximum Values of Ground Motion		
	Acceler- ation	Velocity	Displacement
	9	In/sec	Inch [*]
"Standard" Relative values	0.5	24	18
Typical Maxima:			
EL CENTRO 1940,horiz.	0.33	16	17
EL CENTRO 1940,vert.	0.22	11	8
Minimum, horiz. [†]	0.10	5	4
Minimum, vert. [†]	0.07	3	3
Very Intense Earthq.	0.75	36	27

[*] Transient motion not involving relative fault displacement.

[**] Minimum values recommended for use in design of nuclear reactors in any region, even where earthquakes are not considered probable.

Table 1: Relative Values of Maximum Ground Acceleration, Velocity and Displacement (ref.10)

may be modified to reflect site foundation conditions in a very general sense by multiplying these parameters by site factors. The recommended values of these factors are (i) 0.67 for competent rock; (ii) 1.0 for soft rock and firm sediment; (iii) 1.5 for soft sediment.

Based on an extended series of empirical studies Newmark and Hall [10] found that smooth response spectra could be determined from ground motion spectra simply by multiplying the latter quantities by amplification factors. These factors depend on the fraction of critical damping; amplification factors for various degrees of damping are given in Figure 2, which is reproduced from the literature. [10] Recommended structural damping values, which depend on stress level and type

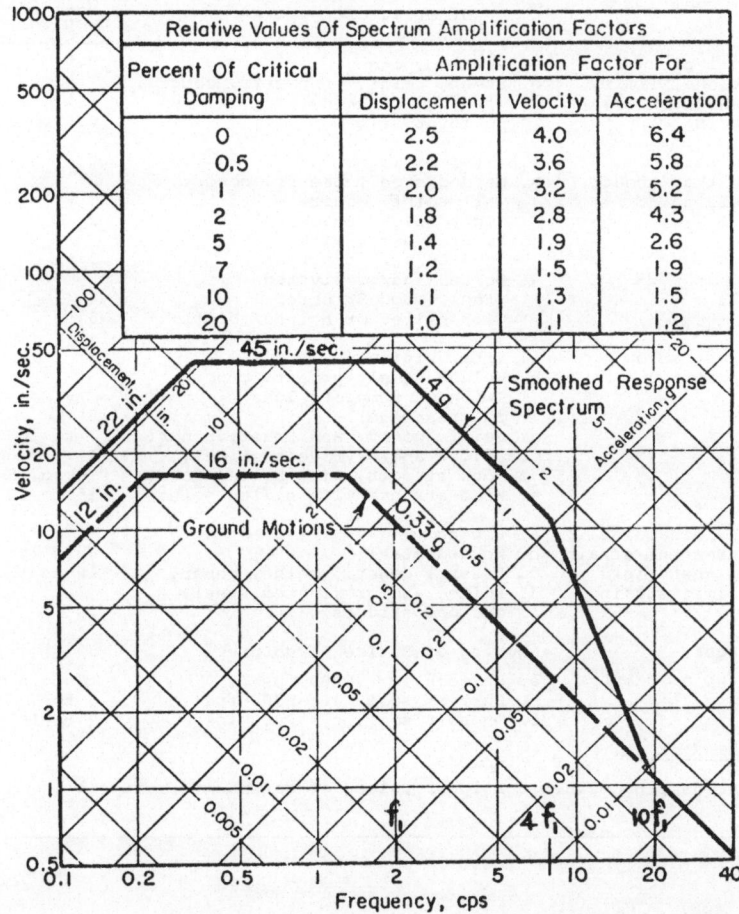

Relative Values Of Spectrum Amplification Factors			
Percent Of Critical Damping	Amplification Factor For		
	Displacement	Velocity	Acceleration
0	2.5	4.0	6.4
0.5	2.2	3.6	5.8
1	2.0	3.2	5.2
2	1.8	2.8	4.3
5	1.4	1.9	2.6
7	1.2	1.5	1.9
10	1.1	1.3	1.5
20	1.0	1.1	1.2

Figure 2:   Smoothed Tripartite Logarithmic Response Spectrum for 0.33 g Earthquake, 2% Critical Damping (ref.10).

of construction, are noted in Table 2. Figure 2 also
shows the constructed forms of the ground and 2%
damped response spectra for a 0.33 g earthquake on
firm sediment. The straight line bounds of the basic
amplified response spectrum are drawn parallel to their
counterpart ground spectrum bounds according to the
above rules. The complete acceleration bound has three
parts. The response spectrum is seen to merge with the
ground motion acceleration bound for high frequency systems. The transition to this state is accomplished
through a set of straight line acceleration branches
intersecting at f_1, $4f_1$ and $10f_1$, where f_1, is the
corner frequency at which the velocity and initial acceleration response bounds intersect.

Stress Level	Type and Condition of Structure	Percentage of Critical Damping
1. Low, well below proportional limit, stresses below 1/4 yield point	a. Vital piping b. Steel, reinf. or prestr. concr., wood; no cracking, no joint slip	0.5 0.5 to 1.0
2. Working stress, no more than about 1/2 yield point	a. Vital piping b. Welded steel, prestr. concr., well reinf. concr. (only slight cracking) c. Reinf. concr. with considerable cracking d. Bolted and/or riveted steel, wood Structs with nailed or bolted joints	0.5 to 1.0 2 3 to 5 5 to 7
3. At or just below yield point	a. Vital piping b. Welded steel, prestr. concr. (without complete loss in prestress) c. Reinf. concr. and prestr.concr. d. Bolted and/or riveted steel, wood structs, with bolted joints e. Wood structs with nailed joints	2 5 7 to 10 10 to 15 15 to 20
4. Beyond yield point, with permanent strain greater than yield point limit strain	a. Piping b. Welded steel c. Prestr. concr., reinf. concr. d. Bolted and/or riveted steel, or wood structs	5 7 to 10 10 to 15 20
5. All ranges	Rocking of Entire Structure [*] a. On rock, c > 6000 fps b. On firm soil, c $\geq$ 2000 fps c. On soft soil, c $\leq$ 2000 fps	 to to 5 5 to 7 7 to 10

[*] Higher damping values for lower values of seismic velocity c .

Table 2: Damping Values (ref.10)

Adjustments to the various spectra shape described above should be considered when designing for structures having unusual dynamic characteristics (involving differences from the norm upon which the preceding curves have been formulated) or for ground motions exhibiting special features (such as motions with relatively larger proportions of high frequency content). The necessary modifications can be achieved by altering or emphasizing certain regions of the spectra to incorporate the unusual features of ground and system characteristics. [11]

EXTRAPOLATED AND ARTIFICIAL ACCELEROGRAMS

The seismic input can be expressed as an acceleration-time hisrory record rather than as a design response spectrum. This becomes a requirement if a non-linear response analysis is contemplated. Once the maximum probable and possible earthquake characteristics are defined from the hazard study, appropriate strong motion records, measured under comparable conditions and meeting the threat specifications, can be selected from the catalogue of existing accelerograms and used in a dynamic analysis. Since the number of recorded motions is still sparse, available records can be modified, in the manner described by Seed, Idriss and Kiefer [9] to provide an accelerogram exhibiting properties (amplitude, frequency content and duration) which are consistent with the expected M or MM intensity and causative fault distance defined by the seismic risk study. The modification is achieved by scaling with respect to design ground acceleration amplitude (ordinate scale) and/or predominant period (time scale). Composite accelerograms can be assembled in this manner when it is necessary to obtain a record with a long duration of strong shaking, such as is exhibited by an earthquake of large magnitude (long fault break).

Because of the incomplete sampling of recorded earthquakes, various authors [12] have constructed artificial earthquakes with known statistical properties for use as design earthquakes. These synthetic events are generated by certain stochastic processes so as to produce the frequency content, amplitude variation and shaking durations exhibited by their prototype motions. A simulated record should be scaled so as to produce a relatively continuous response spectrum which matches the smooth design spectrum postulated for a site. Definitions of four artificially generated motions are briefly noted in an accompanying paper.[1]

When the acceleration-time history approach is followed, a family of different inputs should be used in the analysis to account for the probabilistic nature of the earthquake record. The design should be based on the maximum response derived from the ensemble analysis.

FINAL REMARKS

Figure 3, taken from a paper by Cornell,[13] is used to summarily illustrate the preceding discussions. It shows design response spectra and the peak response

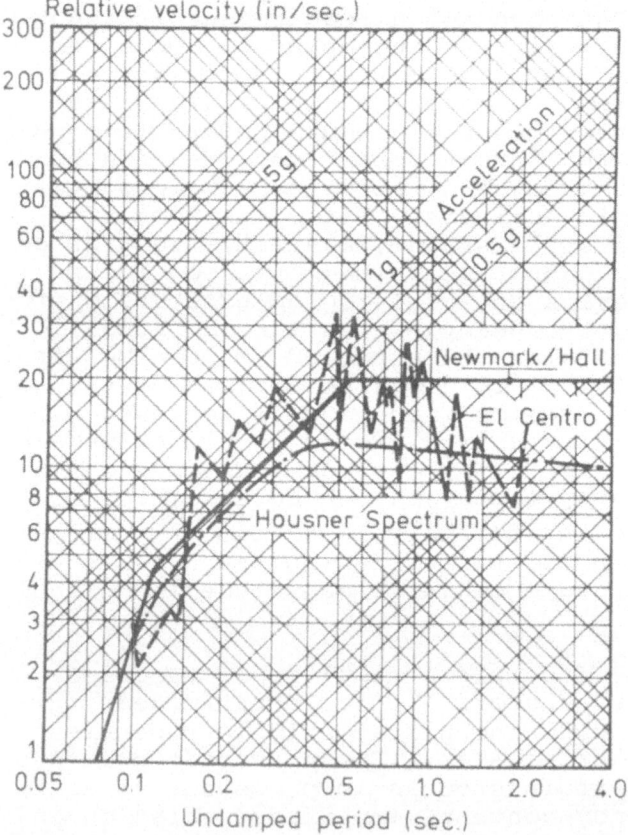

Figure 3: Design Response Spectra and Peak Response Due to Modified El Centro, NS,1940,(ref.13).

due to a catalogue earthquake, which was modified to represent a 0.1g ground acceleration input for the site. El Centro, NS 1940, whose peak acceleration amplitude was 0.33 g , was chosen as a catalogue record which could be suitably extrapolated to a design earthquake; its ordinate was accordingly scaled by a 0.1 g/0.33 g factor. The frequency content and duration of shaking for this earthquake were assumed to conform with the corresponding properties of the desired design motion. As expected, the design spectrum based on enveloped values (Newmark) lies above the spectrum associated with average response values (Housner) . The actual earthquake forces determined by these spectra would depend on the allowable design stresses specified with each curve. The normalized El Centro response lies considerably above the average design spectrum for most periods, indicating the severity of this record in relation to the average.

The general approach for the seismic risk determination of a specific site can be extended to construct areal seismic zoning maps. Design response spectra and design accelerograms can be specified for each zone, using the principles outlined above.

Once the design inputs are chosen, or defined, the calculations for structural response can proceed by following standard dynamic analysis techniques.

REFERENCES

1 Cherry, S., Earthquake Ground Motions: Measurement and Characteristics, NATO Advanced Study Inst. in Eng. Seism. and Earthq. Eng., Izmir, Turkey 1973.

2 Gumbel, E.J., Extreme Value Statistics, Columbia U. Press, N.Y., 1958.

3 Milne, W.G. and Davenport, A.G., Distribution of earthquake risk in Canada, Bull. S.S.A., Vol 59, 754, April 1969.

4 Cornell, C.A. and Vanmarcke, E.H., The major influences of seismic risk, 4th World Conf. Earthq. Eng., Chile 1969.

5 Dick, J.D., Extreme Value Theory and Earthquakes, 3rd World Conf. Earthq. Eng., New Zealand 1965.

6 Housner, G.W., *Earthquake Engineering*, Prentice Hall, 1970.

7 Algermissen, S.T., Seismic risk studies in the United States, *4th World Conf. Earthq. Eng.*, II, 865, Chile 1969.

8 Neumann, F., A broad formula for estimating earthquake forces in oscillators, *2nd World Conf. Earthq. Eng.*, Japan, 1960.

9 Seed, H.B., Idriss, I.M. and Kiefer, F.W., Characteristics of rock motions during earthquakes, *Rept. No EERC 68-5*, College of Eng., U. of California, Berkeley, 1968.

10 Newmark, N.M. and Hall, W.J., Seismic design criteria for nuclear reactor facilities, *4th World Conf. Earthq. Eng.*, Chile, 1969.

11 Fallgren, R.B., Jennings, P.C. and Smith, J.L., Aseismic design criteria for electrical facilities, *Preprint 2014, ASCE Nat. Struct. Eng. Mtg.*, San Francisco, 1973.

12 Jennings, P.C., Housner, G.W. and Tsai, N.G., Simulated earthquakes for design purposes, *4th World Conf. Earthq. Eng.*, Chile, 1969.

13 Cornell, A.C., Seismic design for nuclear power plants, M.I.T. Press, 114, 1970.

PROBLEMS IN SEISMIC ZONING

by   G.W.Housner

Professor of Civil Engineering and Applied
Mechanics, California, Institute og Technology, Pasadena, California.

and   P.C.Jennings

Professor of Applied Mechanics,
California Institute of Technology,
Pasadena, California

SYNOPSIS

The paper analyzes the problems of seismic zoning
for engineering design and makes recommendations for
future practice. The degree of data and judgement that
are present in seismic zoning maps of different types
is reviewed and the characteristics of the different
forms of data are examined. A brief section is devoted
to an assessment of micro-zoning and the theoretical
calculation of surface motions.

It is concluded that a good seismic zoning map for
engineering use should be simple, with broad zones, and
should not be overly dependent on individual past earthquakes. In the writers' judgment, seismic zoning maps
specifying design criteria should be drawn by knowledgeable engineers in each particular field, using more
general scientific maps for data and guidance. It is
concluded also that microzoning and the theoretical
calculation of surface motions are not yet reliable
methods for determining ground motions for design calculations; it is more appropriate to determine such
motions by direct extrapolation from comparable recorded accelerograms.

INTRODUCTION

A seismic zoning map for engineering use is a map
that specifies the levels of forces or motions for
earthquake-resistant design, and thus it differs from
a seismicity map that provides information only about
the occurrence of earthquakes. Seismic zoning maps are
practical tools in earthquake-resistant design because
they provide useful guidance when it is not feasible to
make thorough studies of the earthquake hazard at particular locations. Such studies can, in general, only
be justified for large projects such as major dams,
nuclear power plants, etc. It must be realized, however, that the forces specified by a seismic zoning map
are only a part of the overall earthquake-resistant design procedure; to assess the true level of earthquake
resistance implied by a design technique it is necessary to know the allowable stresses, strains, deflections,
damping, ductility, etc., used in the design process.

The construction of seismic zoning maps is made
difficult by the lack of adequate data, by the conflict
between the needs for safety and economy, and by a lack
of knowledge of the occurrence of earthquakes and the
detailed character of potentially damaging earthquake
motions. This lack of important knowledge is, of
course, one of the major reasons for having seismic
zoning maps, as it is unreasonable under these conditions to expect engineers specializing in design to
make judgments about the earthquake hazards of a particular site. It is better for experts to apply the required judgment by making zoning maps. A map representing the compilation of the best available data and
judgment of knowledgeable geologists, seismologists,
earthquake engineers and design engineers is a practical
necessity.

The undertaking of individual assessments of earthquake hazards at particular sites is an alternative to
the use of a zoning map which is logically attractive
and which has been used, for example, in California for
determining recommended design forces for school buildings. Such investigations are not as thorough as those
required for a nuclear power plant or a major dam and
the results tend to be markedly inconsistent because of
the combination of lack of basic data and different
viewpoints and backgrounds of those making the assessments. A seismic zoning map that is based on all avail-

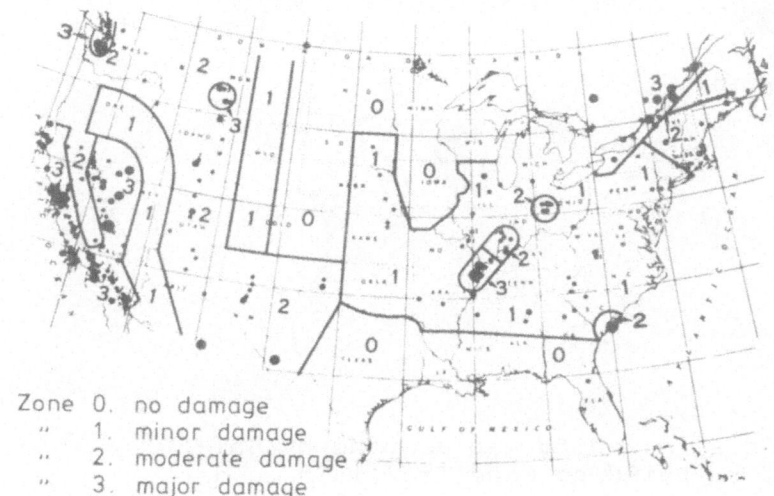

Zone 0. no damage
 "   1. minor damage
 "   2. moderate damage
 "   3. major damage

Figure 1:   Seismic zoning map adopted by Uni-
            form Building Code, 1949-1970.

able information and which considers consistently the
various features for a particular application would be
better for the public as well as the design engineer.

    The first seismic zoning map used in the United
States was compiled by the U.S.Coast and Geodetic Survey. in 1948 [1], revised in 1940,[2] (reflecting the 1949
earthquake near Olympia, Washington, and downgrading
the region near Charleston, South Carolina from Zone 3
to Zone 2), and officially withdrawn in 1952.[3] The
1949 map, shown in Figure 1, was adopted, however, into
the Uniform Building Code in 1949 and was retained
there until 1970,[4] when it was replaced by the revised
"Seismic Risk Map" developed by ESSA/Coast and Geodetic
Survey in 1969,[5] shown in Figure 2. The maps now used
by the Uniform Building Code (UBC) of the United States
and the National Building Code of Canada are shown together in Figure 3. The Maps are for the same purpose,
but have been prepared by different groups. The maps
are thought to be fair examples of the state of seismic
zoning. Although the maps match at their common boundary in broad features, it is clear that seismic zoning,
particularly in areas of moderate to low seismicity, is
not yet a precise science.

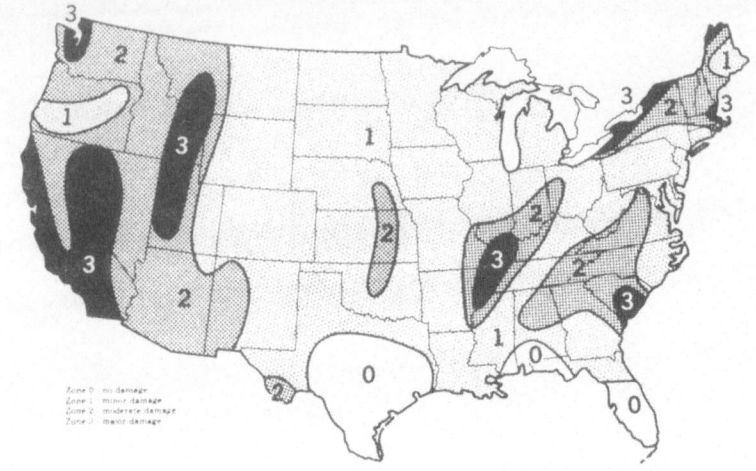

Figure 2:   Seismic zoning map adopted by Uniform Building Code from 1970 to date.

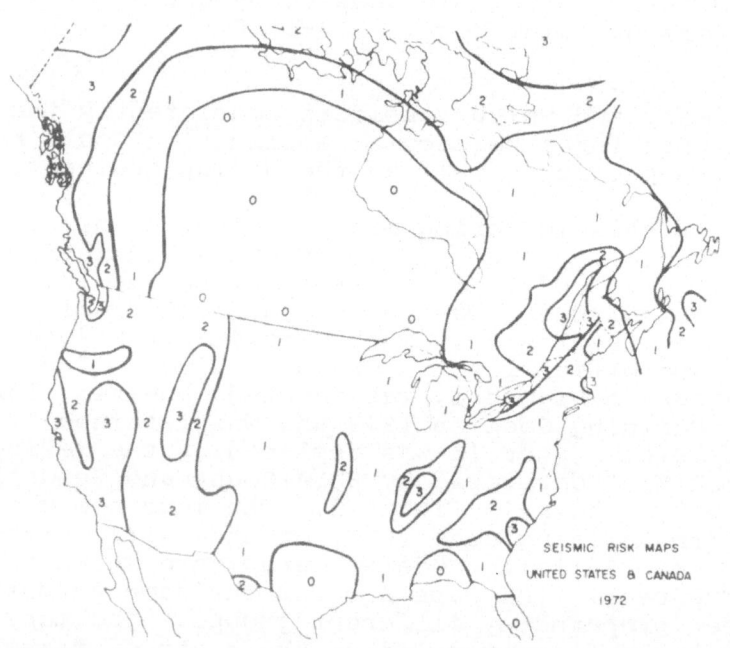

Figure 3:   Comparison of seismic zoning maps now
in use in building codes of Canada and U.S.

The methods used to prepare both of the maps in
Figure 3 are explained in papers published in the Proceedings of the 4WCEE. They are constructed on different principles, for example, the frequency of occurrence of earthquakes was not included as a factor in
preparing the map adopted by the UBC, even though this
is a very important factor for engineering purposes.

TYPES OF SEISMIC ZONING MAPS

A wide variety of seismicity maps, seismic zoning
maps, etc. have been prepared, depending on whether the
purpose of the map is to present basic data, to present
data augmented by judgment, or to specify criteria for
design; and it is important to distinguish among these
different maps. Within the last category there can be
major differences depending on the application for
which a map was prepared, as well as on the relative
emphasis given to economy and safety. These maps can
be categorized by the extent to which they portray a
mixture of data and judgment.

Seismicity Maps

The simplest seismicity map is a plot of magnitude
-rated epicenters of historical earthquakes and jugd-
ment enters only to the extent of interpreting the
pre-instrumental history of the region, if included.
Epicentral locations alone can be quite misleading, for
they do not give information on the areal extent of
strong shaking for large earthquakes. Because of this,
seismicity maps are sometimes drawn on the basis of the
Modified Mercalli Intensity Scale, or similar scales.
Although such maps give a better indication of the
areal effects of strong shaking, the interpretation of
the intensity scales is highly judgmental, and maps
based on intensities contain a much higher ratio of
judgment to data than maps of epicenters. For regions
in which the recorded seismic data is scanty, the Modified Mercalli "epicentral intensity" is often used in
place of magnitude to rate epicenters plotted on a map,
and such maps reflect the imprecision associated with
the epicentral intensities, which may be very great.

If data were available adequate to define the frequency of occurrence vs. magnitude of earthquakes over
a region, an earthquake probability map could be constructed. In most parts of the world, data is insuf-
ficient for this purpose, and much estimation must be

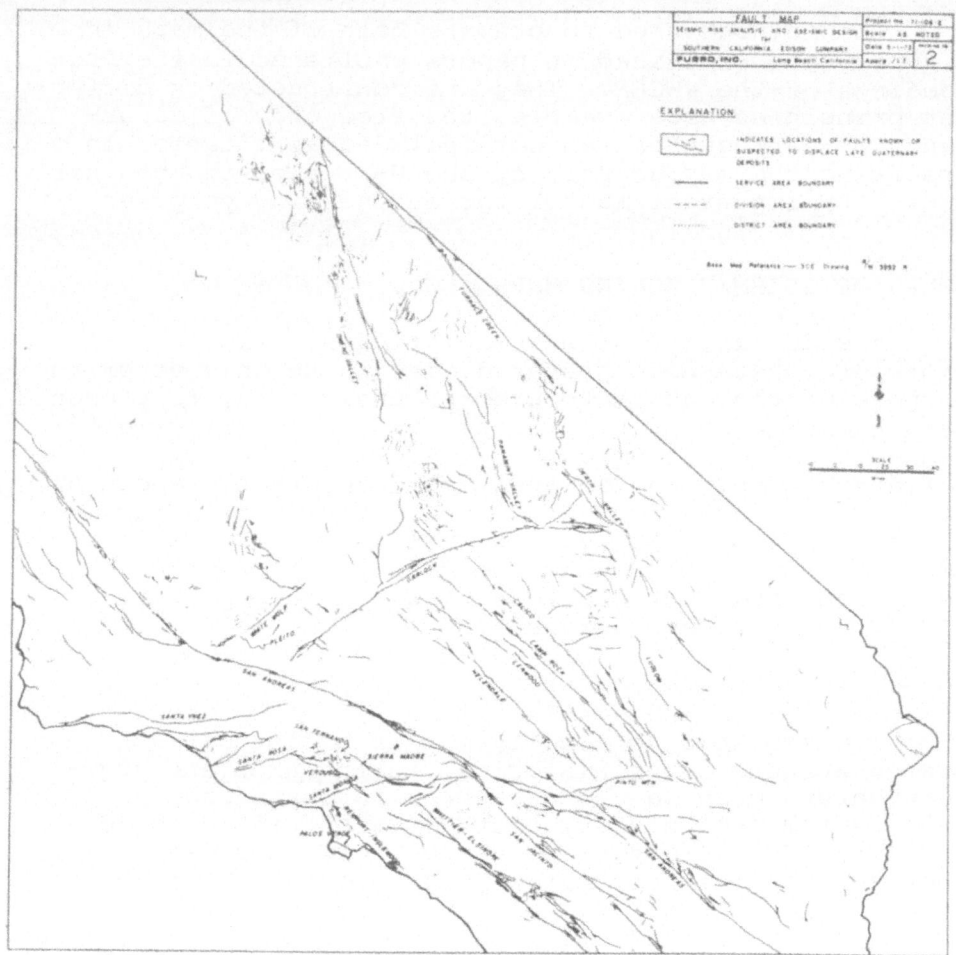

Figure 4: Fault map of southern California showing
           faults judged to be active within past
           500,00 years. [6]

employed.

An alternate method of presenting seismicity data
that is sometimes used is to plot the distribution of
strain energy density associated with historical earthquakes, resulting, for example, in a contour map of cumulative strain-energy density. The disadvantage of
maps of this type from an engineering viewpoint, is
that individual earthquakes are not identified, and the

smooth variation of the contour plots may be misleading
as to future ground shaking to be expected.

Fault Maps, Seismotectonic Maps and Seismic Probability
Maps.

Seismic fault maps are intended to show all faults
on which movement has taken place within certain specified periods of time, e.g., in historic times or in
the last 10,000 years . An example of a map of this
type is shown for southern California in Figure 4.[6] A
great deal of judgment is involved in the preparation
of such a map, even for historic earthquakes in readily
accessible locations. Differences of professional
opinion arise over the length of fault rupture, the
differentation of faulting from effects of slides, and
over the inferred characteristics of subsurface faults.
The difficulties are similar in the assessment of
faults that have moved within recent geologic time, but
the differences of professional judgment can arise over
more major features and may revolve, for example, over
whether a certain fault is active in the prescribed
sense. Even if the fault is judged to be active, the
probability of earthquakes of various magnitudes occur-
ing on the fault can only be estimated imprecisely, so
that the use of a fault map must be based on a very
large measure of judgment.

We use the name "seismotectonic map" to describe a
map which is essentially a fault map augmented by other
geologic information, such as inferred tectonic processes, local geology, etc. There is a considerable
variety in the information that can be included in such
maps and there is clearly a major component of judgment
in the selection and presentation of the material.

A seismic probability map is sometimes constructed
from a fault map by assigning probabilities of occurrence of earthquakes of different magnitudes to each
active fault, and assigning areal distributions of intensity to earthquakes of different magnitudes. By
this procedure a map is constructed that shows a probability of intensity of shaking. Large components of
judgment and estimation are required to construct such
a map.

Engineering Maps

The previously discussed maps present basic data
combined with professional seismological and geologic-

Figure 5: Seismic zoning map prepared by Fugro, Inc.
for Southern California Edison Co.
Intended use is for earthquake-resistant
design of electrical equipment.[6]

al judgment, and are generally not directly useful to
the engineers, whose need is for quantitative guidance
regarding seismic loads to be resisted within certain
allowable stresses, strains, etc. Maps prepared for
engineering use can vary significantly depending both
upon the intended applications and the interests of the
individuals constructing the map. For example,a seismic
zoning map for high-rise buildings might differ from

one appropriate for short-period structures; a seismic zoning map for nuclear power plants might differ from one for single-family dwellings; and a map for use in southern California might logically differ in character from a map for use in India. Also, the need for some equipment to be portable or the requirements of manufacture may dictate a different type of seismic zoning map for special applications. (An example might be equipment for elevators, for which uniformity of manufacture would suggest a single seismic zone for the entire western United States. An example of an engineering zoning map intended for electrical equipment in southern California is shown in Figure 5.[6] The different zones in the map are keyed to different levels of response spectra and scaled accelerograms.

Dissimilarities in engineering zoning maps arising from differing interests of the individuals preparing them are usually the result of compromises between the requirements of economy and public safety. A map made solely from the viewpoint of public safety would be based on the strongest credible shaking, and the frequency of occurrence of the shaking would not be a factor. On the other hand, a map zoned on the economics of providing earthquake resistance vs. the cost of repair would be based on the probability of strong shaking, and no special design for earthquakes would be required in zones where the return period of potentially damaging shaking was significantly longer than the average life structures, regardless of the hazard to public safety should such shaking occur. It is easily recognized that most existing maps, including the U.S. map in Figure 3 and its predecessors, are compromises between these two extremes. In effect, a seismic zoning map is usually a reflection of the fact that economic considerations prevent the provision of the more complete protection otherwise desirable.

The construction of a map for engineering use clearly requires major input of judgment beyond that already embodied in the seismological and geological maps upon which they are based. We also note that factors other than technical matters have a legitimate place in the construction of seismic zoning maps for engineering purposes. Jurisdictional boundaries which affect the administration of the zoning map may suggest convenient locations of borders of zones that are ill-defined by technical information. In fact, the oddly shaped boundaries of the seismic zones in the map in Figure 1 were employed to emphasize the lack of precision in the data on which the map was based.

RELIABILITY OF THE BASIC DATA

The most important basic data for earthquake-resistant design are the strong-motion accelerograms recorded at many sites throughout the world. An eventu-
al goal for seismic zoning would be a map keyed to col-
lections of strong-motion records sufficient to define
the seismic loading in each zone. Unfortunately, the
data do not yet permit this, and this goal will not be
reached for many years. The present collection of data
is sufficient to produce samples of strong shaking from
major earthquakes recorded under a variety of conditions, and is adequate for defining probabilities of
occurrence for large areas such as California and Japan.
There is not, however, enough data to define with precision the motion expected at a site or the frequency
of occurrence of earthquakes in small area, nor has the
strong shaking been recorded in a great, Magnitude 8+
earthquake.

~The data represented by historic accounts of earthquake damage and by seismological measurements are one
step removed from the problem because it is necessary,
lacking strong-motion data, to infer the strength and
character of the shaking. In the case of seismological
measurements it is necessary to use some relation between a seismological measure of the earthquake, such
as Richter Magnitude, and engineering measures of the
strength of shaking, such as spectrum intensity.

Although Intensity Scales, such as Modified Mercalli, are intended to measure the severity of shaking,
they are not defined to do this effectively. The most
serious weakness is a failure to separate effects on
engineered structures from general effects on people,
objects and nonengineered construction. The higher
ratings of these scales are also well known to be overly sensitive to effects exhibited by soils which can
occur under a wide range of severity of shaking. In
the light of current information it appears that a
given MM Intensity rating can be produced by shaking of
various strengths and durations, and the Intensity ratings of a given earthquake cannot be accepted as indicative of the severity of shaking in an engineering
sense without further examination.

Empirical relations between Richter Magnitude and
the level of strong shaking are more reliable, but the
most common of these, which relate Magnitude and epicentral distance to maximum acceleration, are known

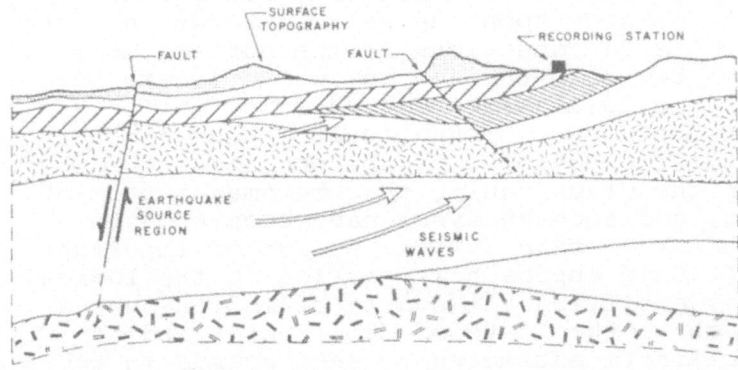

Figure 6: Schematic diagram showing relation
of earthquake source, travel path,
and local site conditions.

from comparisons with data to be very approximate. For
example, it was observed in the San Fernando earthquake
that the peak accelerations at locations equally distant
varied by a factor of two.[7] Furthermore, by definition,
the Magnitude is the maximum response in the long-
period range, measured at distances of 100 km or more,
whereas the peak acceleration is in the short-period
range and is measured at distances as short as a few
kilometers. It is not surprising, then, that these
quantities cannot be precisely related by an empirical
equation.

Use of evidence of faulting in geologically recent
but pre-historic times, and other tectonic information,
is another step further removed from the problem. Judgment is required to estimate the possible occurrence of
earthquakes from such data and also to infer the Magnitude of potential earthquakes. These judgments are
often stated as additions to the historic record in
terms of Magnitude or Intensities, thus reducing the
engineering problem to that discussed above.

MICROZONING AND CALCULATIONS OF SURFACE MOTIONS

Several methods have recently been proposed to
calculate surface motions in earthquakes, given the occurrence of an earthquake of specified Magnitude on a
specific fault. As shown in Figure 6, the main feature
from the point of view of calculating surface motions

are the nature of the source mechanism, the effect of
the travel path geology upon the seismic waves, and the
effects of local site conditions on the surface motion.
Some techniques begin at the source, while others begin
at the base of the local site, with bedrock motion adjusted for the distance from the fault. When combined
with a probabilistic estimate of the occurrence of
earthquakes on the given fault, this becomes a form of
seismic zoning, and such analyses have been used to
estimate earthquake motion for the design of important
facilities. If this approach is carried to its logical
conclusion by identifying all source of motion over a
large area, such as the State of California, and at-
taching probabilistic estimates to each source in a
consistent manner, the results could be compared with
the historic record to assess the accuracy of the process. This has not yet been done, and it is our feel-
ing that most studies of this type would, if done for a
large area, result in probabilistic estimates for strong
shaking that are significantly higher than can be sub-
stantiated by historic seismicity.

Efforts to calculate surface motions by beginning
somewhere in the source-travel path-site effects chain
are useful scientific studies that have as their goal
the explanation of the nature of earthquake motions.
Only the surface motions have been measured in strong
earthquakes, however, and therefore it is not yet possible to distinguish the effects of assumptions about
the different portions of the chain. The assumptions
required are quite severe and, in our opinion, can in-
volve errors exceeding those attendant on directly
estimating the surface motion from extrapolation of recorded motions. Furthermore, strong-motion records obtained at El Centro [8] indicate that for firm soils,
local site effects are less important than effects of
different source mechanisms and travel paths. Similarly, the motions measured in Pasadena during the San
Fernando earthquake showed behavior inconsistent with
the usual results of calculation techniques. [9] Such results, and the present lack of measured bedrock and
source motions indicate that these methods are over
-simplified, and are not at this time capable of reliably calculating the surface motion in most practical
cases.

The use of microtremors and microseisms for seismic zoning is also a valid topic for scientific research, but does not yet appear to be a reliable method
of seismic zoning for purposes of engineering. The

principal difficulties are the observed nonstationarity of the motions, [8] the unknown character of the sources of the motions, and the lack of correlation with the measured characteristics of strong ground shaking.

CONCLUSIONS AND RECOMMENDATIONS

Characteristics of a Good Map

A major feature of a good seismic zoning map is that it should not change substantially after the occurrence of an earthquake: this means that, if possible, the map should not be overly dependent on any one past earthquake. This ì the weakness of maps formally derived from a history which is too short to establish the seismicity of the region under study. In a practical sense, the users of a seismic zoning map really want to know what earthquake ground shaking their structures will experience. Thus, it is unavoidable that the quality of a seismic zoning map is to a large degree dependent on how well it predicts the future occurrence of ground shaking.

Another major characteristic of a good seismic map is that it not be overly elaborate. In our opinion, it is a mistake to try to draw highly detailed maps or to put a lot of information on the map. Such refinements are a barrier to understanding by those who would use the map. A map intending to specify criteria for earthquake engineering design must do so unambiguously so the user will not have to make further judgments that are out of his area of competence. A map giving expected MM Intensities or a map showing expected maximum ground acceleration are examples of maps in a form not directly useful to the engineers.

A good seismic map for use in engineering design should be adapted to the particular needs and design practice of the users. Thus, it is to be expected that a seismic zoning map for tall buildings might differ in important details from one to be used for electrical transmission facilities. It is not good practice, in general, to adopt a map and associated design criteria that were developed for a particular application directly into a different application.

Ground Motions for Design

It is our judgment that the most appropriate way

to select earthquake motions for purposes of design is
to assemble a group of strong-motion records recorded
under as comparable conditions as possible, and to extrapolate from these records, by simple scaling, the
required motions. The records used can be augmented
by artificial earthquake records if necessary. This
simple approach seems best suited to the present state
of knowledge, and makes clear to potential users the
nature of the judgments involved. The more elaborate
approaches that introduce additional approximations and
estimations without providing any more basic data
tend to obscure the distinction between information derived from reliable data and the results provided by
approximate methods of calculation.

Construction of Seismic Zoning Maps

In agreement with other authors, it is our conclusion that a number of maps should be drawn with each
group of workers preparing the type of map that is
within its area of competence. For example, government
agencies are the logical groups to prepare large-scale
maps of seismicity, faulting, and other pertinent geologic features. Such maps would form the basic data
for seismic zoning maps to be used to specify forces
for design. It is important that maps specifying criteria for design be updated periodically. Such updat-
ing is required because of expected increases in the
knowledge of earthquake effects, and because the earthquake protection demanded by society changes with time,
increasing directly with increasing urbanization and
industrialization.

In view of the state of knowledge in the various
disciplines that contribute to seismic zoning for engineering design, and the large judgmental factors involved, it appears to us that the only workable way to
develop a seismic zoning map for the purpose of specifying criteria for design in a particular field is to
convene a group of the most knowledgeable people in the
field and have them construct the map. The group would
have to review first the scientific maps and data,with
advice as needed, and assess to its best judgment the
degree of conservatism embodied in current design procedures. A seismic design map drawn by such a group
should represent a reasonable balance of the various
factors that are important to the earthquake-resistant
design problems of their segment of the engineering.

REFERENCES

1. Roberts, E.B., F.P.Ulrich, "Seismological Activi-
    ties of the U.S. Coast & Geodetic Survey in 1948",
    BSSA, 40, 195-216, 1950.

2. Roberts, E.B. and F.P.Ulrich, "Seismological Acti-
    vities of the U.S. Coast & Geodetic Survey in
    1949," BSSA, 41, 205-220, 1951.

3. Roberts, E.B. and F.P.Ulrich, "Seismological Acti-
    vities of the U.S. Coast & Geodetic Survey in
    1951," BSSA, 43, 3, 255-268, 1953.

4. International Conference of Building Officials,
    "Uniform Building Code, 1970 Edition, Volume I",
    Pasadena, Calif. 1970.

5. Algermissen, S.T., "Seismic Risk Studies in the
    United States," Proc. Fourth World Conf. on Earth-
    quake Eng., Vol. I, Santiago, Chile, 1969.

6. Fugro, Inc., "Seismic Risk Analysis and Aseismic
    Design Manual," prepared for Southern Calif. Edi-
    son Co., May, 1972.

7. Hudson, D.E. (Ed.), "Strong-Motion Instrumental Da-
    ta on the San Fernando Earthquake of Feb. 9, 1971",
    Earthquake Eng. Research Lab., Calif. Inst., and
    Seismological Field Survey, NOAA.

8. Udwadia, F.E., "Investigations of Earthquake and
    Microtremor Ground Motions," Earthquake Eng. Re-
    search Lab. Report 72-02, Calif. Inst. of Techn.,
    Pasadena, Calif, Sept. 1972.

9. Hudson, D.E., "Local Distribution of Strong Earth-
    quake Ground Motion," BSSA, 62, 6, Dec. 1972.

FUNDAMENTALS OF DYNAMIC EARTHQUAKE RESPONSE ANALYSIS

by Julius Solnes

University of Iceland, Reykjavik, Iceland.

INTRODUCTION

When subjected to earthquakes, man-made structures will mainly have to withstand a forced horizontal shaking motion at their foundations. Depending on the severity of the earthquake motion in the vicinity of the structure, the induced vibrations in the structure may be entirely within the elastic limits of the structure, whereby the only results are possible discomfort of inhabitants and minor damage of secondary elements. However, more than often, the induced earthquake vibrations in the structure will take it well into the nonlinear or inelastic range, resulting in permanent damage of the structure. In even moderate earthquakes, it is likely that slight structural damage has to be tolerated in order to conform with economically feasible design practice. Finally, in strong-motion earthquakes, severe failure or total collapse is to be expected because of large plastic deformations in the structure.

Figure 1 shows a typical earthquake acceleration, i.e. that of Taft, California 1952,[1] which may be imposed upon the base of the structure. It is evident that the response of the structure can only be described properly through a very complicated dynamic analysis in which the vibratory characteristics of the structure, the structure-foundation-soil system and the earthquake motion are all important.

Before the advent of electronic computers, any attempt to assess the dynamic behaviour of complex structures in earthquakes appeared to be futile. It was only possible to assert that the earthquake acceleration at the base would correspond to fictitious inertia forces acting horizontally on the structure. In an earthquake of magnitude $M \sim 7.0$, the maximum horizontal ground acceleration to be expected may be of the order of one third the acceleration of gravity. Neglecting the frequency dependent amplification of the structural response, a statically equivalent force of $1/3\ mg$, applied horizontally at the centre of gravity of a structure with mass m, would produce the same maximum effect in the structure, i.e. maximum deflection, maximum base shear etc. Therefore, the notion of percent gravity design was early adopted in Japan and the USA and gradually implemented in the building codes of most earthquake countries. The famous $0.1\ g$ earthquake design rule may still be used in many countries.

Inadequate as the percent gravity rule may be, it has still proved satisfactory in many cases, and at least it has ensured a certain minimum horizontal force resistance for the structures so designed, necessary to resist lateral forces of any kind. In many earthquakes, however, it has been observed that structural damage and failure are quite haphazard so that similar structures of equal percent gravity design may experience vastly different amounts of damage or destruction. Thus after a destructive earthquake, two identical structures of the same design, a few miles apart, the one may stand totally undamaged and the other be totally destroyed. Clearly the answer to this implied question can only be given through detailed dynamic analysis of the structures.

The earthquake record in Figure 1 shows a frequency content corresponding to wave periods in the range $0 < T < 6$ sec. Almost all civil engineering structures will have a fundamental natural period of vibration lying in this same period interval. Now, the initial 5-30 seconds destructive motion of almost all recorded earthquakes show a similar frequency composition to the Taft earthquake, biased towards the high or low portion of the above period range. Therefore, earthquake damage is mainly due to quasiresonant earthquake response or intense response amplification in the overlapping frequency ranges.

Analysis of near field earthquake data will reveal

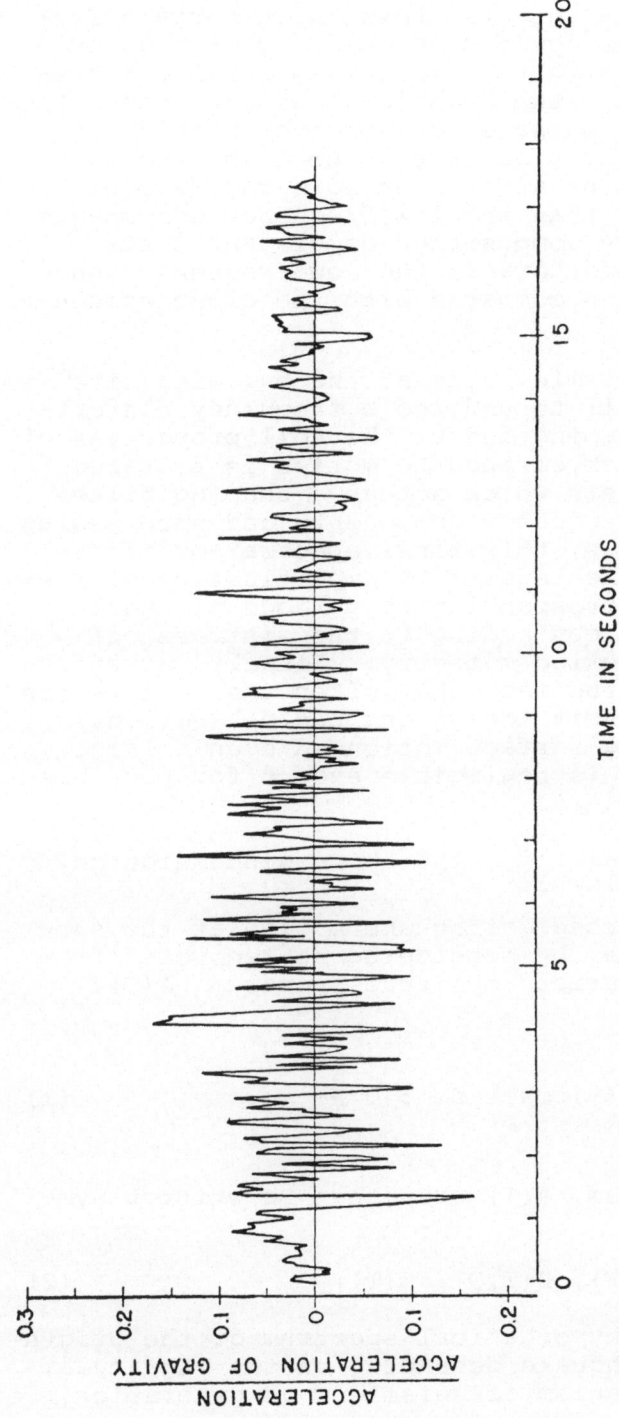

Figure 1. TAFT July 21, 1952.Accelerogram S 69 E.

that the earthquake waves are "born" with certain frequency content at the focus such that higher magnitude
earthquakes will be slightly biased towards lower frequencies. [2] Moreover, the seismic waves are attenuated
and distorted during passage through the earth's crust
to the location of the structure. The high frequency
wave components will be attenuated more rapidly, with
increasing distance, than the low frequency components,
whereby the frequency composition of distant earthquakes will be shifted towards the low frequency range
or higher periods when compared with the close epicentral earthquakes. [2,3]

The incoming seismic waves at the building site
considered, have still to undergo a frequency distortion and amplitude shaping due to the soil properties of
the site. The bedrock earthquake motion is filtered
through the soil layers which act as a shaping filter
amplifying certain frequency components and suppressing
others. In many cases, this final spectral modification of the earthquake results in amplification of frequency components corresponding to periods of about
0.2-2.0 seconds, thereby adding to the distress of
structures in this period category. Finally, attention
should be called to the fact that often the soil layers
are too weak to transmit the strong bedrock motion,
whereby the resulting surface motion in even a large
magnitude earthquake is negligible except for possible
large fault breaks. [4]

Following K.Kanai [5,6] , the above discussion may
be summarized as follows:

Denoting the undisturbed earthquake motion at the surface by x(t) , it may be considered as recovered from
the corresponding Fourier amplitude spectrum A(T),
i.e.

$$x(t) = \frac{1}{\pi} \int_0^\infty A(\omega)\cos\omega t \, d\omega \; ; \; \omega = \frac{2\pi}{T} \tag{1}$$

The amplitude spectrum A(T) can then be written sym-
bolically as

$$A(T) = F\left[O(T), \, D(T,Q), \, H(T)\right] \tag{2}$$

where O(T) is the hypothetical spectrum of the origin
or focus of the earthquake depending on the magnitude,
focal depth and mechanism of seismic wave generation.

D(T,Q)   represents the vibration characteristics of the
earth's crust, i.e. the distortion, refraction and reflection, wave absorption and attenuation of the travelling seismic waves, and   H(T)   describes the amplification or subsoil characteristics of the site.   Q   is
the specific attenuation factor. [7]

Assuming now that a simple structure is subjected
to the earthquake motion   x(t)   at its base, the response of the structure   y(t)   can formally be written
as:

$$y(t) = \int_{-\infty}^{\infty} x(s)h(t-s)ds \qquad (3)$$

The response   y(t)   can obviously be considered as yet
another interpretation of the original earthquake o(t).
The impulse response function   h(t)   of the structure
is but a new shaping filter corresponding to the vibratory characteristics of the structure.   Hence, the
final amplitude response spectrum of the structure, by
which the maximum stress and strain in the structure is
obtained, can be written as

$$Y(T) = HS(T) \cdot F\Big[O(T),D(T,Q),H(T)\Big] \qquad (4)$$

where   HS(T)   is the frequency response function of the
structure, i.e.

$$HS(T) = \int_{0}^{\infty} h(s)\cos\frac{2\pi}{T}s \ ds \qquad (5)$$

In the next sections, the response analysis of
earthquake-excited structures is studied with regard to
the above introduced concepts.   The results of such
analysis are discussed and a foundation is laid for the
introduction of earthquake design loads.

EARTHQUAKE RESPONSE ANALYSIS

Consider a simple structure of the inverted pendulum type subjected to an earthquake acceleration $\ddot{x}(t)$
at its base, (Fig.2).   If the structure can be represented as a simple one-degree-of-freedom oscillator,
the equation of motion is given by

$$m(\ddot{y}(t) + \ddot{x}(t)) + c\dot{y}(t) + ky(t) = o \tag{6}$$

where m is the mass, k is the spring constant and c the viscous damping coefficient. y(t) is the deflection of the mass relative to the base. Upon reduction, Equation (6) is written as

$$\ddot{y}(t) + 2\lambda\omega_o\dot{y}(t) + \omega_o^2 y(t) = -\ddot{x}(t) \tag{7}$$

where ω_o is the natural frequency and λ the ratio of critical damping.

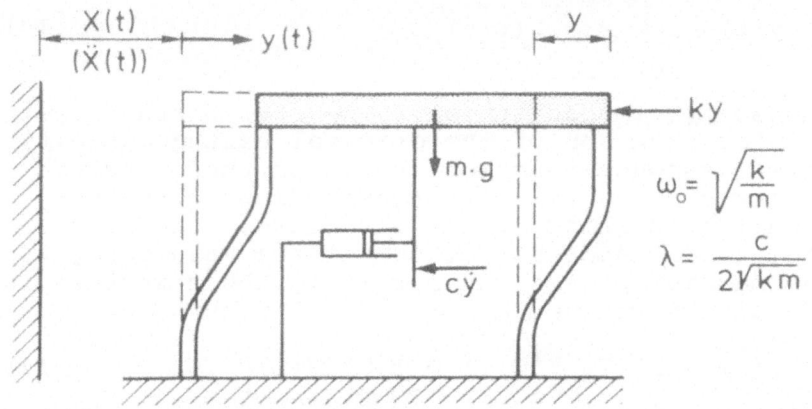

Figure 2. Single-degree-of-freedom structure.

Now, the response of the structure y(t) may firstly be considered as recovered from the amplitude spectrum (see (1)) or

$$y(t) = \frac{1}{2\pi} \int_{-\infty}^{\infty} Y(\omega)e^{i\omega t}\, d\omega \leftrightarrow Y(\omega) = \int_{-\infty}^{\infty} y(t)e^{-i\omega t}dt \tag{8}$$

in which the complex notation has been adopted, i being the imaginary unit and Y(ω) the complex Fourier amplitude spectrum of the response. Fourier transformation of the entire Equation (7) yields

$$-\omega^2 Y(\omega) + 2\lambda\omega_o\omega i Y(\omega) + \omega_o^2 Y(\omega) = \omega^2 A(\omega) \tag{9}$$

or

$$Y(\omega) = \frac{1}{(\omega_o^2 - \omega^2) + 2i\lambda\omega_o\omega} \cdot \omega^2 A(\omega)$$

$$= H(\omega) \cdot \omega^2 A(\omega) = \int_{-\infty}^{\infty} y(t) e^{-i\omega t} dt \qquad (10)$$

$H(\omega)$ is the complex frequency response of the structure (cf.(5)). Then by (8) and (10),

$$y(t) = \frac{1}{2\pi} \int_{-\infty}^{\infty} H(\omega) \cdot (\omega^2 A(\omega)) e^{i\omega t} d\omega \qquad (11)$$

and finally applying the convolution theorem

$$y(t) = -\int_{-\infty}^{\infty} \ddot{x}(s) h(t-s) ds \qquad (12)$$

(cf. (3)). The impulse response function

$$h(t) = \frac{1}{2\pi} \int_{-\infty}^{\infty} H(\omega) e^{i\omega t} d\omega \qquad (13)$$

and since by (10),

$$H(\omega) = \frac{1}{(\omega_o^2 - \omega^2) + 2\lambda i\omega_o\omega} \qquad (14)$$

the integration in (13) can be carried out to yield ($\lambda < 1$)

$$h(t) = \begin{cases} \dfrac{\exp(-\lambda\omega_o t)}{\omega_o\sqrt{1-\lambda^2}} \sin(\omega_o t\sqrt{1-\lambda^2}), & t \geq o \\ \\ o, & t < o \end{cases} \qquad (15)$$

For most civil engineering structures, the damping ratio λ will be small ($\lambda < 0.10$) whereby the final approximate solution in the time domain, ((12) and (15)), is obtained.

$$y(t) = -\frac{T_o}{2\pi} \int_0^t \ddot{x}(s) \left[\exp(\lambda\frac{2\pi}{T_o}(s-t))\sin\frac{2\pi}{T_o}(t-s) \right] ds \qquad (16)$$

The two solutions (10) and (16) give rise to two different interpretations of the response. Firstly, the time domain response (16) will attain a maximum value at a certain time during the earthquake, say t_{max}. Then, the main response parameters of the structure are obtained as follows:

$$\left| Y_{max} \right| = \frac{T_o}{2\pi} \left\{ \int_0^t \ddot{x}(s) \left[\exp(\lambda\frac{2\pi}{T_o}(s-t))\sin\frac{2\pi}{T_o}(t-s) \right] ds \right\}_{t=t_{max}} \qquad (17)$$

$$\dot{y}_{max} = - \left\{ \int_0^t \ddot{x}(s) \left[\exp(\lambda\frac{2\pi}{T_o}(s-t))\cos\frac{2\pi}{T_o}(t-s) \right] ds \right\}_{max} \qquad (18)$$

$$(\ddot{y}+\ddot{x})_{max} = \frac{2\pi}{T_o} \left\{ \int_0^t \ddot{x}(s) \left[\exp(\lambda\frac{2\pi}{T_o}(s-t))\sin\frac{2\pi}{T_o}(t-s) \right] ds \right\}_{max} \qquad (19)$$

Except for the cosine term in the integral expression for the maximum velocity (18), all the three response parameters can be written in terms of the common response parameter, $S_v(\lambda,T)$, denoted by the pseudo-velocity-response,[1] or

$$\left. \begin{aligned} Y_{max} &= \frac{T}{2\pi} S_v(\lambda,T) \\ \dot{y}_{max} &\approx S_v(\lambda,T) \\ (\ddot{y}+\ddot{x})_{max} &= \frac{2\pi}{T} S_v(\lambda,T) \end{aligned} \right\} \qquad (20)$$

The true velocity differs only slightly from the pseudo velocity, hence the name, and for all engineering purposes, the above indicated approximations have been found to be sufficiently accurate. [8]

The concept of the pseudo-velocity spectrum which is defined by the family of çurves obtained for

$$S_v(\lambda,T) = \left[\int_0^t \ddot{x}(s)\{\exp(\tfrac{2\pi}{T}\lambda(s-t))\sin\tfrac{2\pi}{T}(t-s)\}ds\right]_{max} \tag{21}$$

by letting T vary through the range $o<T<3$ seconds for different values of the ratio of critical damping λ , has proved extremely useful in earthquake resistant design.[9] It is easily related to various simple design criteria for a structure the behaviour of which may be approximated by that of a simple linear oscillator.

Thus the maximum strain energy in the structure during the earthquake is

$$E_{max} = \tfrac{1}{2} ky_{max}^2 = \tfrac{1}{2} mS_v^2 \tag{22}$$

The maximum shear force at the base is

$$Q_{max}=ky_{max}=k\tfrac{T}{2\pi}S_v=\tfrac{2\pi}{T}mS_v=m(\ddot{x}+\ddot{y})_{max} \tag{23}$$

The horizontal design load is usually expressed in terms of a seismic coefficient C and the weight of the structure W , i.e.

$$F = CW , \qquad W = mg \tag{24}$$

Then by (23)

$$C = \tfrac{2\pi}{T} \tfrac{1}{g} S_v \tag{25}$$

in order to have the same maximum deflection of the structure.

Figure 3 shows the pseudo-velocity spectrum of the Taft earthquake,[1] , which portrays all the general features of such spectra. A small amount of damping will reduce the spectral ordinates considerably and the curves become almost flat for large periods and increased damping.

Housner[9] has proposed the smoothed spectra shown in figures 4a and 4b as a semiprobabilistic basis for aseismic design. These are obtained through averaging the computed real spectra of eight recorded components of Californian earthquakes. The more significant features are the spectral peak at 0.5 seconds of the un-

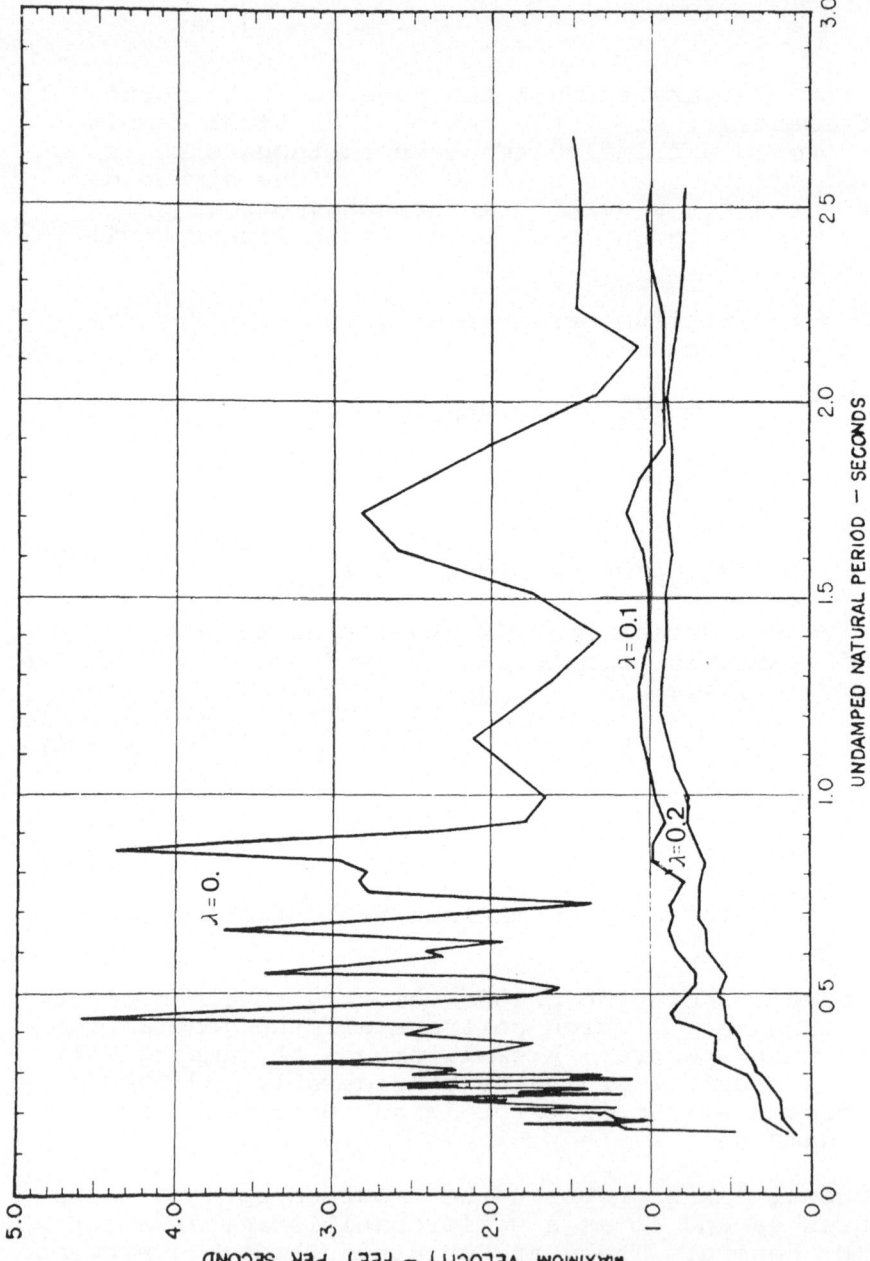

Figure 3. Pseudo velocity spectrum for TAFT, July 21,1952,S 69 E.

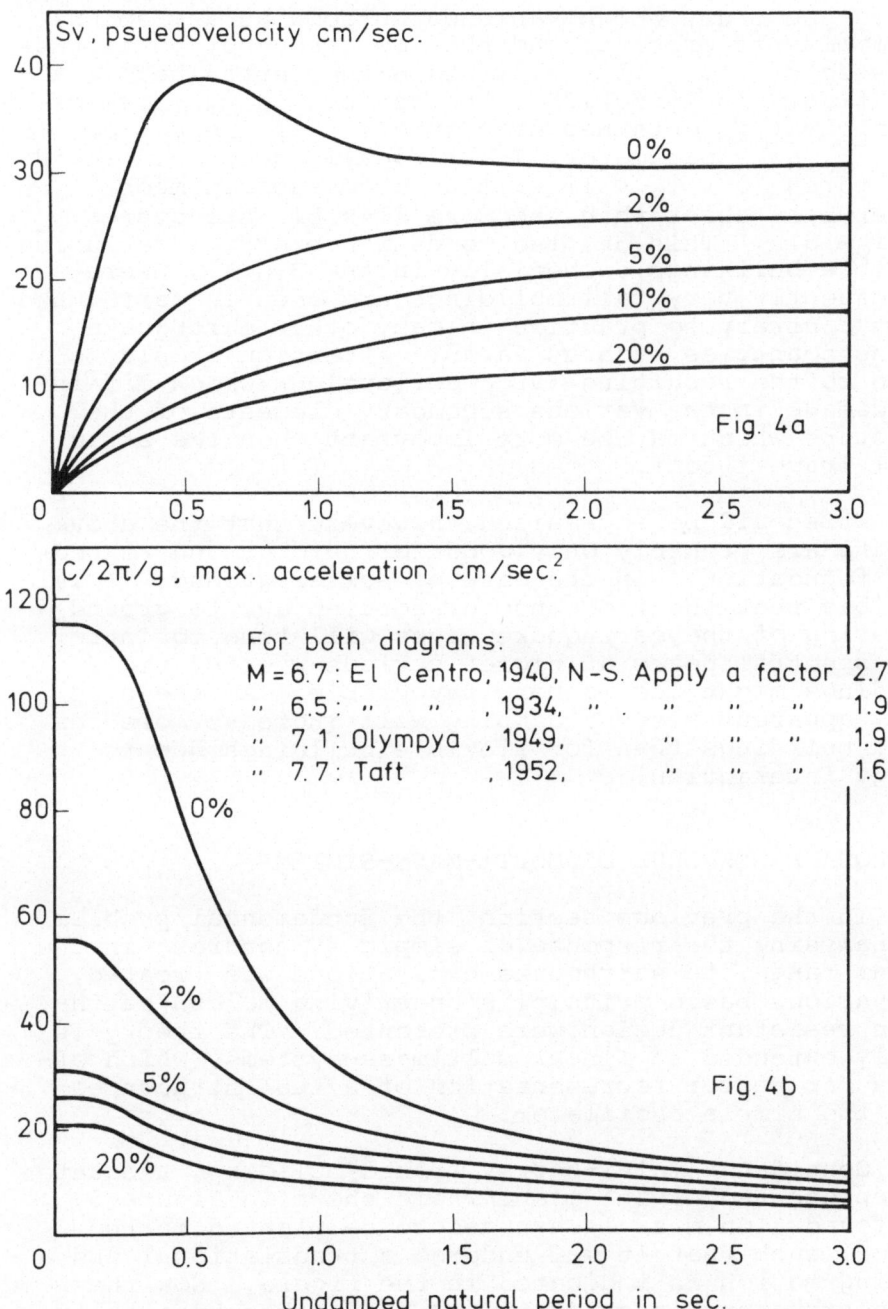

Figure 4. Averaged, smoothed pseudo velocity spectra by G.W. Housner (ref. 9).

damped spectrum, which vanishes as soon as any amount
of damping is present, and the low frequency portion of
the spectra (1.5 < T < 3.0) where the curves become almost flat. Applying (25), the corresponding seismic
coefficient is obtained as a function of the period T
with λ as a parameter. It is clearly seen that rigid
structures (o < T_o < 1) will have to sustain much
larger base shear than the more flexible structures
(1< T_o< 3) . This has led to design practice favouring
flexible buildings, especially in the U.S.A., where
consequently very tall buildings are seen in earthquake
areas contrary to practice in many other earthquake
-prone countries such as Japan. Attention should be
drawn to the resulting large deflections which may in-
cur damage in the various secondary elements of the
structure which may be more important than the structural frame itself.

It should be emphasized, however, that the above
spectra are probably only good for hard ground to bedrock foundation. In the case of softer ground, it is
possible that the frequency distortion due to ground
filtering of the earthquake waves will tend to introduce spectral peaks at higher periods whereby rigid
buildings might become more favourable. At the same
time, apparent viscous damping will increase more for
rigid buildings than for flexible buildings due to
energy interaction.

EARTHQUAKE RESPONSE OF MULTI-MASS-SYSTEMS

In the previous section, the fundamental problem
of analysing the response of simple structures, in the
linear range, to earthquake excitations was treated,
and various basic principles underlying modern earthquake resistant design were presented. The theory is
easily extended to linear multimass-systems, which offer a far better representation of actual structures
than the simple oscillator.

Consider a multi-storey framed building, adequately represented by the plane frame shown in Figure 5.
The foundation mass is assumed to be planted in the
subsoil such that it may undergo a translational and a
rocking motion as indicated in the figure. Now the
whole structure is subjected to an earthquake acceleration $\ddot{x}(t)$ which causes the reference frame of the
structure to undergo the motion x(t) . Let m_i be
the mass concentrated at the i'th floor, defined as the

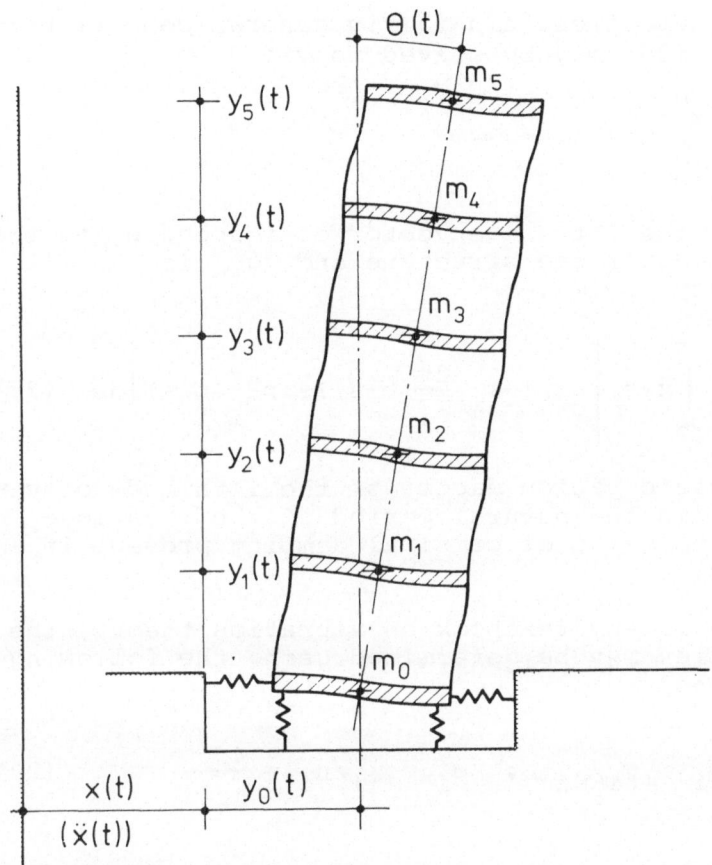

Figure 5. Multi-degree-of-freedom structure.

mass of the i'th floor plus the contributions from half
the adjacent upper and lower storeys. Then the equations of motion for the whole structure, using matrix
notation, may be written as follows $(y_o(t) = \theta(t) = 0)$:

$$\underset{\sim}{M}\ddot{\underset{\sim}{Y}} + \underset{\sim}{C}\dot{\underset{\sim}{Y}} + \underset{\sim}{K}\underset{\sim}{Y} = -\underset{\sim}{M}\underset{\sim}{E}\ddot{x}(t) \quad ^{\dagger)} \tag{26}$$

where $\underset{\sim}{M}$ is the (NxN) diagonal mass matrix, $\underset{\sim}{C}$ is the
(NxN) symmetric matrix of the damping coefficients and
$\underset{\sim}{K}$ is the (NxN) symmetrical stiffness matrix. $\underset{\sim}{E}$ is a
(Nx1) column vector of unit elements.

---

†) In the text, a tilda ~ is used to designate a matrix or a vector.

Under a set of certain fairly general conditions[10] the equations (26) may be solved to yield

$$Y = \sum_{i=1}^{N} V^i q_i A_i \tag{27}$$

where V^i is the i'th eigenvector corresponding to the i'th normal mode of the structure and q_i is the i'th coordinate response, i.e.

$$q_i(t) = -\frac{T_i}{2\pi} \int_0^t \ddot{x}(s) \left[\exp(-\lambda_i \frac{2\pi}{T_i}(t-s)) \sin\frac{2\pi}{T_i}(t-s) \right] ds \tag{28}$$

A_i is the participation factor of the i'th mode of vibration, T_i is the natural period of the i'th mode and λ_i is the ratio of critical damping present in the i'th mode.

As shown in any textbook on vibration theory, the above quantities may be determined using the following valid relations:

$$2\lambda_i \omega_i = \sigma_i/\mu_i \quad ; \quad \omega_i^2 = \chi_i/\mu_i = \frac{4\pi^2}{T_i^2} \tag{29}$$

where

$$(V^i)^T M V^i = \mu_i , \quad (V^i)^T C V^i = \sigma_i , \quad (V^i)^T K V^i = \chi_i \tag{30}$$

and of course,

$$\left. \begin{array}{c} \det|K - \omega_i^2 M| = 0 \\[2ex] (K - \omega_i^2 M) V^i = 0 \end{array} \right\} \tag{31}$$

The participation factor A_i is then given by

$$A_i = ((V^i)^T M E)/((V^i)^T M V^i) \tag{32}$$

or written out, element for element

$$A_i = \frac{\displaystyle\sum_{n=1}^{N} m_n V_n^i}{\displaystyle\sum_{n=1}^{N} m_n (V_n^i)^2} \tag{33}$$

where V_n^i is the modal deflection of the n'th mass in the i'th mode.

Analogous to the frequency domain solution of the simple oscillator (10), a frequency domain solution of Equation (26) is easily obtained.

Multiply all elements in (26) by the exponential factor $\exp(i\omega t)$ and then integrate over the whole frequency range; hence,

$$\underset{\sim}{Y}(\omega) = \underset{\sim}{H}(\omega)(\omega^2 \underset{\sim}{M} \cdot \underset{\sim}{E} \cdot X(\omega)) \tag{34}$$

where $\underset{\sim}{Y}(\omega)$ is the Fourier transform vector of the response vector $\underset{\sim}{Y}(t)$ and $\underset{\sim}{H}(\omega)$ is the complex frequency response matrix

$$\underset{\sim}{H}(\omega) = (-\omega^2 \underset{\sim}{M} + i\omega \underset{\sim}{C} + \underset{\sim}{K})^{-1} \tag{35}$$

As before, the time domain solution is then recovered from the Fourier transform (34), applying the inverse transform (8).

In the case of Equation (26), which by assumption could be uncoupled in N-space to yield the solution (27), the corresponding Fourier transform is also diagonalised or

$$Y_n(\omega) = \sum_{i=1}^{N} V_n^i \frac{\omega^2 A_i}{\omega_i^2 - \omega^2 + 2\lambda_i \omega_i \omega i} X(\omega) \tag{36}$$

where $Y_n(\omega)$ is the Fourier transform of the n'th mass response.

Turning back to the time domain solution (27), it is possible to apply the response spectrum techniques, developed for the linear oscillator, to give an assessment of the maximum response during the earthquake. By (20) and (28), the maximum coordinate response is

$$q_{imax} = \frac{T_i}{2\pi} S_V(\lambda_i, T_i) \tag{37}$$

Now, these maxima do not occur simultaneously so a direct superposition of the maxima will overestimate the true maximum response. However, to avoid complete response analysis requiring precise input data, various proposals have been put forth to give an approximate expression for the maximum response. An upper bound envelope for the maximum deflection of the structure is had by writing

$$\{y_n\}_{max} = \left\{ \sum_{i=1}^{N} |A_i v_n^i| \frac{T_i}{2\pi} S_V(\lambda_i, T_i) \right\} \tag{38}$$

For earthquakes with the power evenly distributed over the frequency range, Rosenblueth[11] has shown that a probable estimate of the maximum response is rather given by the root square sum of the isolated maxima whereby,

$$\{y_n\}_{max} = \left\{ \sum_{i=1}^{N} A_i^2 (v_n^i)^2 \frac{T_i^2}{4\pi^2} S_v^2(\lambda_i, T_i) \right\}^{\frac{1}{2}} \tag{39}$$

Finally, attemps have been made to combine the two formulae (38) and (39) dependent on the flexibility of the structure.[12] If the maximum shear force in the n'th storey is called Q_n,

$$\{Q_n\} = (1-\beta) \left\{ \sum_{i}^{N} |Q_n^i| \right\} + \beta \left\{ \sum_{i}^{N} (Q_n^i)^2 \right\}^{\frac{1}{2}} \tag{40}$$

where Q_n^i is the modal maximum for the shear force in the n'th storey and

$$\beta = \frac{4}{27} \left(\frac{3}{2}\right)^{(\log N/\log 2)} \tag{41}$$

Of interest now, is to determine the acceleration response of the structure in order to evaluate the fictitious inertia forces causing the maximum deflection of the structure during the earthquake. Differentiation of Equation (27), two times, yields ($\lambda_i < 0.10$)

$$\ddot{\underset{\sim}{Y}}(t) + \underset{\sim}{E}\ddot{\underset{\sim}{x}}(t) = \sum_{i=1}^{N} \underset{\sim}{v}^i \frac{4\pi^2}{T_i^2} q_i(t) A_i \qquad (42)$$

Premultiplying by $\underset{\sim}{M}$ gives the corresponding fictitious inertia forces or

$$\underset{\sim}{F}(t) = \underset{\sim}{M}(\ddot{\underset{\sim}{Y}} + \underset{\sim}{E}\ddot{\underset{\sim}{x}}) \qquad (43)$$

The inertia force at the n'th level can now be resolved, element by element, as

$$f_n(t) = \sum_{i=1}^{N} \left(\frac{\sum_{r=1}^{N} W_r v_r^i}{\sum_{r=1}^{N} W_r (v_r^i)^2} \right) \cdot W_n v_n^i \cdot \frac{4\pi^2}{T_i^2} q_i(t) \frac{1}{g} \qquad (44)$$

where $W_n = m_n g$ is the weight concentrated at the n'th floor.

   Equation (44) may serve as a basis for a set of suitable design loads. Evidently the statically equivalent design loads should give a similar maximum deflection of the structure if linear behaviour is adopted. Using the maximum (38), the design loads $\{D_n\}$ are

$$D_n = \sum_{i=1}^{N} \left| \frac{\sum_{r=1}^{N} W_r v_r^i}{\sum_{r=1}^{N} W_r (v_r^i)^2} \right| \cdot W_n v_n^i \, C(\lambda_i, T_i) \qquad (45)$$

where $C(\lambda_i, T_i)$ is the generalised seismic coefficient (cf. (25) and Figure 4b),

$$C(\lambda_i, T_i) = \frac{2\pi}{T_i} \frac{1}{g} S_v(\lambda_i, T_i) \qquad (46)$$

The participation factors $A = \{A_i\}$, are a set of numbers the magnitude of which is rapidly decreasing with increasing modal number. As seen by Equation (33) the first factor A_1 is much the largest since all the elements of $\underset{\sim}{V}^1$ are positive, whereas already A_2 is considerably smaller since about one half the elements of $\underset{\sim}{V}^2$ are negative. Essentially, the bulk of the response information is carried by the first mode only and the 2nd and 3rd modes only slightly modify the result, the remainder being insignificant. Therefore, it is convenient to rewrite (45) in terms of the modified first mode only. A fair approximation is given by the following design formula in which the modified first modal vector is taken to be proportional to the height of each floor, h_n, above ground.

$$D_n = \frac{\left(\sum_{r=1}^{N} W_r h_r \right) \cdot W_n h_n}{\sum_{r=1}^{N} W_r h_r^2} \cdot C(\lambda_1 T_1) \tag{47}$$

The earthquake resistant design of a multi-storey structure is therefore effectuated by providing adequate shear strength in each storey corresponding to the storey design shear forces caused by the design loads D. A convenient design parameter is the base shear force of the structure, Q_B, which is obtained by summing the design loads (47) or

$$Q_B = \sum_{n=1}^{'N} D_n = \frac{\left(\sum_{r=1}^{N} W_r h_r \right)^2}{\sum_{r=1}^{N} W_r h_r^2} \cdot C(\lambda_1, T_1) \tag{48}$$

Then the design loads can be expressed in terms of this parameter,

$$D_n = Q_B \frac{W_n h_n}{\sum_{r=1}^{N} W_r h_r} \tag{49}$$

Finally, it is customary, in the various codes, to specify the design base shear Q_B as a function of the total weight of the structure $W = \Sigma w_r$ (plus a code-specified fraction of the live loads) and a design earthquake spectrum $C(T)$,

$$Q_B = W \cdot C(T) \tag{50}$$

whereby a simple design procedure (50), (49) is formulated.

The above formulae (49) and (50) have been adopted for the Californian aseismic code. [13] They have the merit of being comparatively simple to use and adaptable to modification due to nonlinear effects and accumulation of knowledge in connection with the factor C.

The above discussion has deliberately been kept within the scope of linear structures. However, the main earthquake design problem is to limit the nonlinear deformations in the structure and yet ensure sufficient ductility in the structural frame. These problems will be discussed in other lectures, and it suffices to say that the elastic design obtained by applying (47), for values of $C(\lambda_i, T_i)$ corresponding to only small to moderate earthquakes, is likely to be adequate in stronger earthquakes as well. The thus designed structure will be able to absorb the excess earthquake energy through plastic hysteresis action much more effectively than through elastic deformation. However, the ductility required ($\mu \sim 4$) is often a major design problem, especially in reinforced concrete frames and shear wall structures.

DESIGN LOAD SPECTRA

Intelligent application of the design loads (47) requires a suitable formulation of the generalised seismic coefficient $C(\lambda_1, T_1)$ to be laid down in the aseismic code in effect. [13] Obviously, the main features to be displayed through the seismic coefficient are the spectral characteristics of the response and the inherent damping of the structure during vibration.

Regarding the damping, it has been tacitly assumed that the damping is of the viscous type represented by the fraction of critical damping in each mode. It is beyond the scope of these notes to give a full discussion of the damping problem in its entire complexity (see Solnes,[14] section 3.6). In passing, the following interesting points may nevertheless be noted. During an earthquake, a structure will never be exposed to the full intensity of the surface motion since a fraction of the earthquake energy is refused by the structure in terms of reflected waves. Moreover, a fraction of the kinetic energy generated in the structure is fed back to the ground as refracted waves. In all, it can be shown [15] that this energy loss can be represented by apparent, equivalent viscous damping which is likely to amount to more than two thirds of the overall damping

value. This is particularly pronounced in the case of
rigid structures, and this effect will also be increased in soft soils. In the case of flexible structures on hard ground, this effect is almost negligible.

The following guidelines may be given [14] in order
to select appropriate damping values for various types
of structures. Beginning with tall flexible structures
on hard ground, critical damping ratios as low as 0.01
are to be expected. The damping ratio may then be assumed to increase with increasing rigidity and increasing softness of the ground. Thus a maximum value of
$\lambda = 0.10$ is likely in the case of a rigid structure on
a very soft ground.

The above damping ratios will also depend on the
amplitude or strain level in the structure caused by
the input excitation. If the above damping ratios are
representative of the state in small to moderate earthquakes, an increase in λ is noticed with increasing
strain level until a certain limiting value is reached
after which the nonlinear hysteresis action provides
the main source of energy dissipation. For moderate to
strong earthquakes, then, the damping ratio need not be
taken as less than 0.03-0.05 . It should also be noticed that, especially in flexible structures, energy
losses due to cracking of nonstructural elements, hysteretic joints etc. will increase the apparent damping
ratios. This will to some extent remove the above indicated frequency dependence of λ . Finally, it may
be noted that masonry and wooden structures may have
damping ratios of values as high as 40% critical.

In most aseismic codes, [13] the generalised seismic
coefficient is given as a function of the fundamental
period only, so the question of selecting appropriate
damping ratios does not arise. The design seismic coefficients have mostly been modelled after the curves
in Figure 4b and the order of magnitude of C is selected according to what is considered to be "economically feasible" aseismic design. Possibly, the damping
ratio which is the same for all structures according to
the code, is selected, bearing the economical aspects
strongly in mind.

As explained in the previous sections, the order
of magnitude of C will in a very complicated manner
be dependent on the frequency composition of the earthquake motion and the frequency response characteristics
of the structure. By Equations (4), (20) and (25) this
functional dependence can be expressed indicatively as

$$C(T_1) \simeq \frac{2\pi}{g} \frac{1}{T_1} \; HS(T_1)A(T_1) \qquad (51)$$

since it is not unreasonable to assume strong functional dependence between the Fourier amplitude spectra and the pseudo velocity spectra. In fact, it is easily shown that the undamped pseudo velocity spectrum is almost equal to the Fourier amplitude spectrum of the input acceleration.[8]

In Fig.6, the development of the seismic coefficient C for three different hypothetical situations is shown schematically.

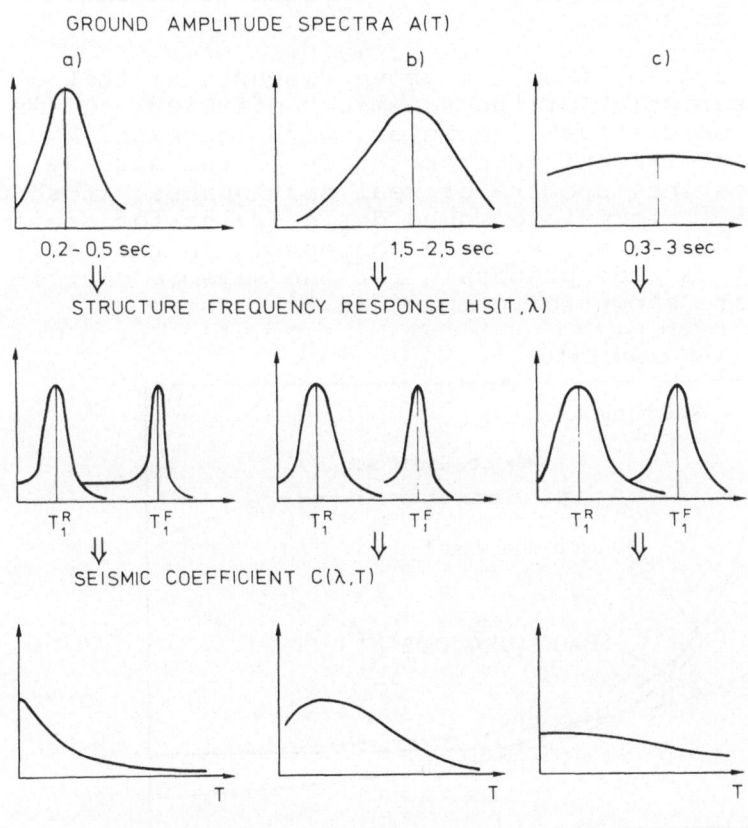

Figure 6. Seismic coefficients C(T) for different soil conditions and type of structure.

Fig.6a covers the development to be expected in the
case of hard ground. The ground amplitude spectrum may
show a distinct peak in the high frequency range and
the corresponding seismic coefficient will have the
same general features as shown in Fig.4b. The damping
ratios of the rigid and the flexible structures respectively, as indicated by the width of the frequency response peaks, will be of a low order of magnitude. The
next case is that of a comparatively soft ground (Fig.
6b). The possible peak in the amplitude spectrum will
be shifted towards the low frequency range and the
damping ratios, especially for the rigid structure,will
be larger than before. Finally, in Fig.6c, a ground
amplitude spectrum is shown which is almost constant at
all frequencies. This could be the situation in the
case of weak soil layers and would lead to seismic coefficients as shown.

It is evident from the above discussion, that a
precise description of the seismic coefficient C is a
fairly impossible task. However, with increased knowledge and intelligent interpretation of the already
existing response spectra of real earthquakes, it should
be possible to formulate sound design rules for aseismic design based on the above concepts. In the final
Fig.7, various code proposals for the seismic coefficient C are shown for comparison.[13]

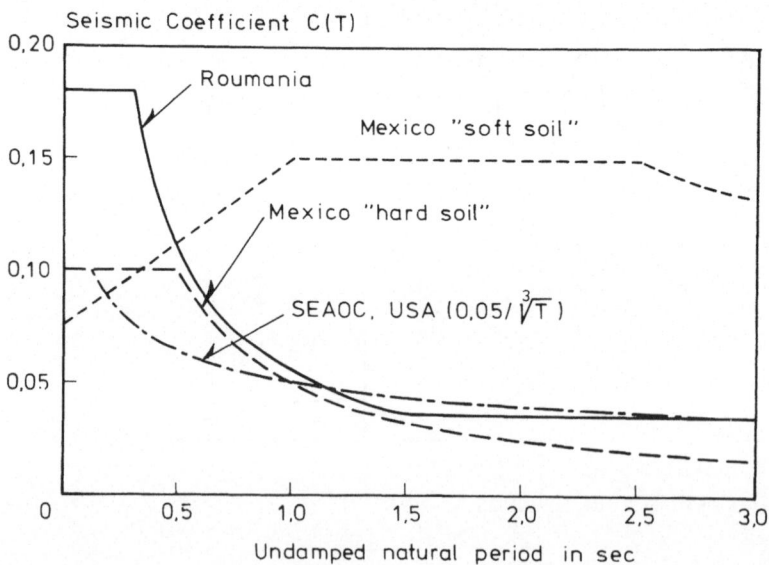

Figure 7. Comparision of design seismic coefficients.

REFERENCES

1. Alford, J.L., Housner, G.W. and Martel, R.R., Spectrum analysis of strong-motion earthquakes, Earthquake Engineering Research Laboratory, California Institute of Technology, Report under ONR, Aug.1951 (revised Aug.1964).

2. Gutenberg, B. and Richter, C.F., Earthquake magnitude, intensity, energy and acceleration (2nd Paper). Bull.Seism. Soc.Amer. 46, 105, 1956.

3. Solnes, J., The Spectral character of earthquake motions, Proc. III.European Symposium on Earthquake Engineering, Sofia, 1, 279, 1970.

4. Ambraseys, N.N. Maximum energy-yield theory, plenary paper, Proc.III.European Symposium on Earthquake Engineering, Sofia, 1, 1970.

5. Kanai, K., Semi-empirical formula for the seismic characteristics of the ground. Bull.Earthquake Research Institute, 35, 307, Tokyo 1957.

6. Kanai, K., An empirical formula for the spectrum of strong earthquake motions. Bull.Earthq. Res. Inst., 39, 85, Tokyo 1961.

7. Knopoff, L., "Q", Reviews of Geophysics, 2, 625, 1964.

8. Hudson, D.E., Some problems in the application of spectrum techniques to strong motion earthquake analysis, Bull. Seism. Soc Amer., 52, 117, 1962.

9. Housner, G.W., Behaviour of structures during earthquakes, Proc. ASCE, Journal of the Eng.Mech. Div., 85, 109, 1959.

10. Caughey, T.K. and O'Kelly, M.E.J., General theory of vibration of damped linear dynamic systems, Dynamics Laboratory, Report, California Institute of Technology, Pasadena 1963.

11. Goodman, L.E., Rosenblueth, E., and Newmark N.M., Aseismic design of firmly founded elastic structures, Transactions ASCE, Paper No.2762, 182,1955.

12. Clough, R.W., Earthquake analysis by response spectrum superposition, <u>Bull. Seism. Soc. Amer.</u>, 52, 647, 1962.

    Arias, S. and Husid, L.R., Response spectrum superposition. A discussion of R.Clough's paper. <u>Bull. Seism. Soc. Amer.</u>, 53, 693, 1963.

13. <u>Earthquake Resistant Regulations. A World List 1970.</u> International Association for Earthquake Engineering, Tokyo 1970.

14. Solnes, E.Julius. <u>Structural Vibration Induced by Earthquakes</u>, Ph.D thesis, Copenhagen 1965.

15. Kanai, K., Tanaka, T. & Suzuki, T., Rocking and elastic vibration of actual building.

    I. Experiments by vibration generator. <u>Bull. Earthq. Research Inst</u>. 36, 183, 1958.

    II. Observation of earthquake motion. <u>Bull. Earthquake Research Inst</u>. 36, 201, 1958.

EARTHQUAKE RESISTANT DESIGN OF TALL BUILDINGS IN JAPAN

by Kiyoshi Muto

Prof.Emeritus, University of Tokyo
Executive Vice-President, Kajima Corporation
President, Muto Institute of Structural
Mechanics

INTRODUCTION

Japanese Building Standard Law, since its establishment in 1921, had prohibited construction of buildings over 31 metres high. However, owing to the extremely high cost of land in urban districts, a movement arose during the 1950's to permit construction of taller buildings for reurbanization purposes and to make more effective use of land. Thus, the repeal of the building height control and revisions of building bylaws were put into effect in 1963. The revisions provide that the conventional law and code apply to the design of common low buildings, whereas the design of tall buildings over 45 metres in height may be carried out by any rational method such as the dynamic design method. No law or code is in force regarding provisions for safety of buildings over 45 metres in height against earthquakes for the time being, in order not to hamper the promotion of research and development in seismic design and construction methods.

In fact the design and construction of tall buildings have to be given final special approval by the Minister of Construction as the precedented cases. For the Minister's approval, the safety of buildings against severe earthquakes must be assured by dynamic analysis by each engineer. Design criteria and judgement on the structural safety, although they are deemed to be reviewed individually by the Examination Board

of the Ministry of Construction, must also be made by
the engineer himself.

It is, therefore, difficult to discuss the present
state of seismic design of tall buildings in Japan as a
whole. The author shows the general features of seismic design which have the most reliable feed-back system, after which he wishes to explain in detail the
design criteria and many links in the feed-back design
system used at the Muto Institute. All the procedures
in an actual design, and the safety estimation are also
shown by the example of the Shinjuku Mitsui building
that will be the highest in Japan.

SEISMIC LOAD IN JAPAN

Seismic Coefficient Method

A structural design method considering seismic
force induced in the building structure during earthquakes was established quantitatively in 1916 by Professor Sano. After the Kanto earthquake in 1923, it
was adopted as the lateral force requirement for the
Building Code. Seismic force regulated in the present
code, which covers all buildings under 31 m with conventional types of materials and techniques, is based
on a seismic coefficient method, that is percent G
method. Member stress due to seismic force has to be
considered together with stress due to fixed and live
load, and be within the allowable limit.

Seismic force f_i is a product of seismic coefficient k_i and weight W_i located at i'th floor. And
seismic coefficient consists of fundamental seismic coefficient, zoning factor and soil-construction factor
(Figure 1).

Basic coefficient k_o is given as 0.2 plus increment i as shown in the Figure 1.The increment i
is defined as 0.01 for every 4 metres over the 4th
floor, or for each floor exceeding 16 metres in the
height of the building. The zoning factor Z is also
shown in the figure, where three grades of the factor,
1.0, 0.9, 0.8, are given according to the seismicity in
Japan. Soil and construction factor S has four grades
of 0.6, 0.8, 1.0 and 1.5 . For instance, the building
with ductile frame on the rock adopts 0.6 while the
building with wall on soft soil does 1.5 as S factor.

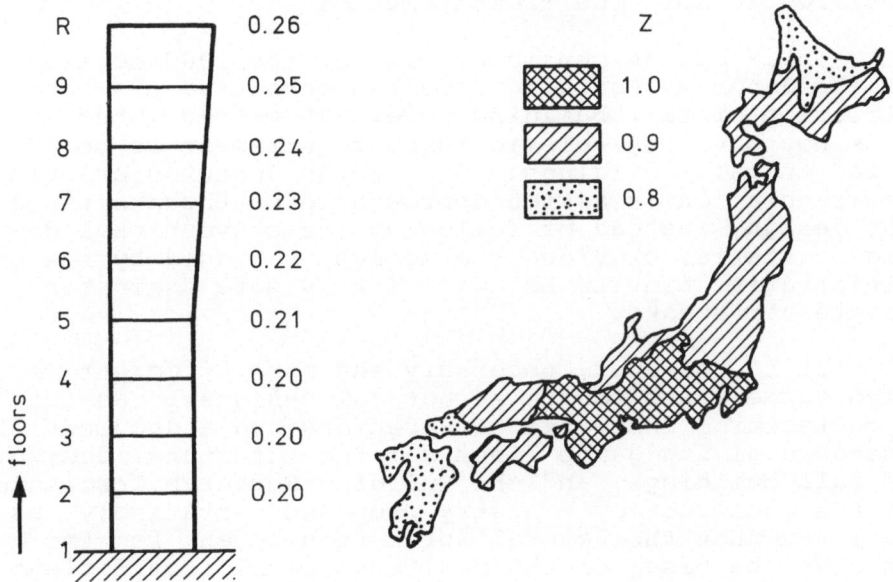

Fundamental Seismic Coefficients Ko   Zoning factor Z

BUILDING CODE FOR H $\overset{<}{=}$ 45 m

Seismic Force   fi = kWi
         where   k = seismic coefficient
                  Wi = i'th floor lumped load

Seismic coefficient

        k = Ko·Z·S
    where Ko = 0.2   for   H $\overset{<}{=}$ 16 M
          Ko = 0.2+i   for   H > 16 M
          i = increment = 0.01 every 4M over 16M
          Z = zoning factor = 1.0, 0.9, 0.8
          S = soil and construction factor
            = 0.6, 0.8, 1.0, 1.5
            (rock → soft soil)

Figure 1: Seismic Loads in Japan (Horizontal Coeffi-
           cients); H ≤ 45 M

Modified Seismic Coefficient Method

In Japan, as mentioned before, the 1963 revision
of the Japanese Building Code has permitted the construction of tall buildings over 45 metres , but it
does not give any seismic force requirement to be applied to tall buildings. One of the features of this
new regulation is a free approach to earthquake resistant design, instead of following the conventional design criteria, provided the design is judged by the administrative body to be sufficiently safe against a
severe earthquake.

It is, however, necessary and more efficient to
give guide lines to the structural designer. So the
Architectural Institute of Japan drafted a document of
standard giving guide lines for the structural design
of tall buildings, and the Building Research Institute
of the Construction Ministry proposed tentatively, as
of 1969, that the lateral force requirement for the
standard be based on the modified seismic coefficient
method.

According to the tentative AIJ-standard, seismic
base shear force is given as a product of base shear
coefficient and total weight of the structure. And base
shear coefficient consists of basic shear coefficient,
zoning factor and importance factor.

Basic shear coefficient C_o is given according to
the fundamental natural period of the building T and
ground condition factor G . Four kinds of G-values ,
-0.75, 0, 0.5 and 0.75, are given as shown in Figure 2
corresponding to rock, gravel, clay and fill respectively. Those concepts are reflected by the research and
investigations into the response spectra of many earthquakes at various sites. But, falling short of our ex-
expectations, it is likely to be conventional that the
maximum values of C_o are limited to 0.2 .

The AIJ-standard also gives the seismic force distribution through the height of the building, which is
the combination of three patterns. The type of pattern
also depends on the fundamental period of the structure.
These patterns are due to uniform, triangular and top
concentrated loading, then lateral seismic force f_i
against i'th floor is defined. Figure 2 also shows
two examples of combined forces in case of the fundamental period of 1.0 sec. and 5.0 sec.

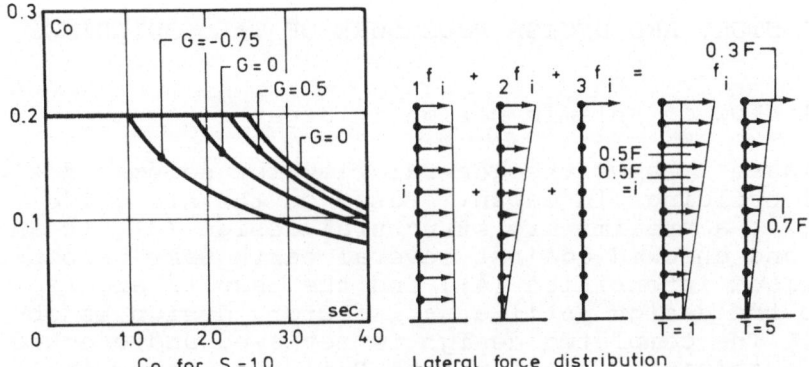

Co for S = 1.0 · · · Lateral force distribution

BRI STANDARD FOR H > 45 M

Seismic Force $F = C_1 \cdot W$
$\qquad\qquad\quad C_1 = Co \cdot Z \cdot I$
where $\qquad\quad$ Co = Basic Shear Coefficient
$\qquad\qquad\qquad$ Co = 0.2S for $\quad$ T<G+1.75 sec

$\qquad\qquad\qquad\quad = \dfrac{0.35S}{T-G}$ for $\quad$ $T \geq G+1.75$ sec

$\qquad\qquad$ T = fundamental natural period
$\qquad\qquad$ G = soil factor
$\qquad\qquad\quad = -0.75, \quad 0, \quad 0.5, \quad 0.75$

$\qquad\qquad$ S = construction factor
$\qquad\qquad\quad = 0.9, \; 1.0$

$\qquad$ Z = zoning factor
$\qquad\quad = 1.0, \; 0.9, \; 0.8$
$\qquad$ I = importance factor

Lateral Seismic Force

$$fi = {}_1fi + {}_2fi + {}_3fi$$
$${}_1fi = \alpha C_1 Wx$$

$$ {}_2fi = \beta C \, Wx \; \frac{W \cdot x}{\displaystyle\sum_{x=0}^{h} Wx \cdot x} $$

$${}_3fi = \gamma C_1 W$$

where

Wx = i'th floor lumped load;
$\;$W = total load;
$\;$x = height of i'th floor;
$\;\alpha = (2-T)/2 \, (T<2), 0 \, (T \overset{\geq}{=} 2);$
$\;\beta = T/2 \, (T<2), \quad (12-T)/10 \, (T \overset{\geq}{=} 2);$
$\;\gamma = 0 \, (T<2), \quad\;\; (T-2)/10 \, (T \overset{\geq}{=} 2)$

Figure 2: Modified Seismic Coefficient According to
$\qquad\qquad$ BRI Standards; H>45 M .

SEISMIC STUDY AND DESIGN PROCEDURE OF TALL BUILDINGS

General Flow of Seismic Design in Japan

Figure 3 shows the general flow for seismic design of tall buildings in Japan. Based on the AIJ guide lines (2), a preliminary structural design (3), is analyzed and checked against several earthquake responses for various intensities (4), and the results are fed back to the design until a satisfactory design is obtained. The completed design is reviewed and checked by the examination board in the Building Center in Japan (5). If the examination board assures the safety of the structure, the Ministry of Construction gives approval for construction.

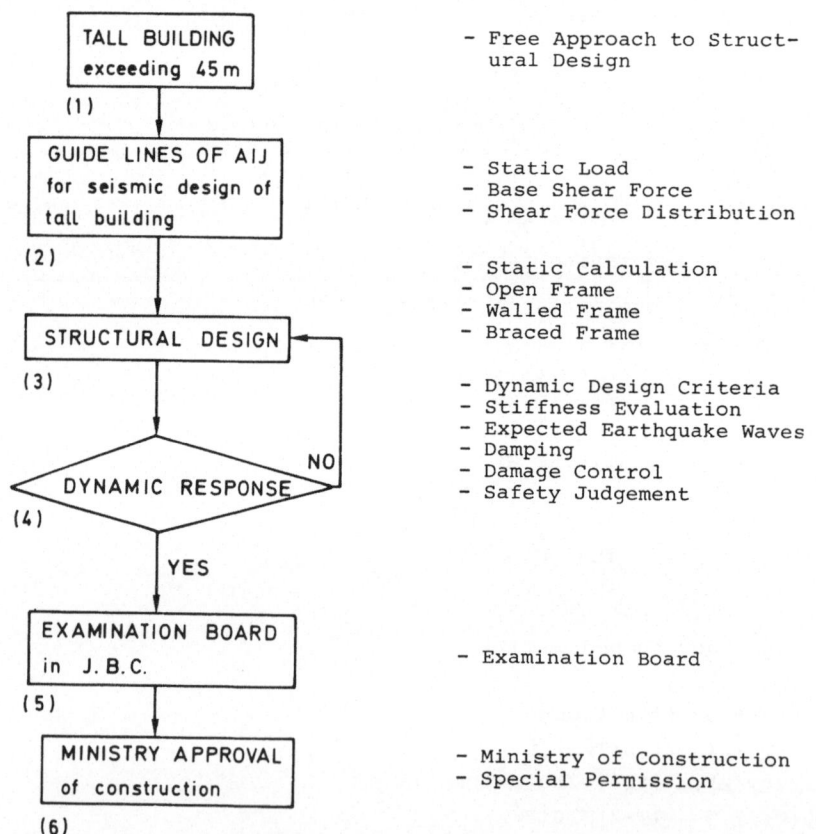

TALL BUILDING exceeding 45 m (1)
- Free Approach to Structural Design

GUIDE LINES OF AIJ for seismic design of tall building (2)
- Static Load
- Base Shear Force
- Shear Force Distribution

STRUCTURAL DESIGN (3)
- Static Calculation
- Open Frame
- Walled Frame
- Braced Frame

DYNAMIC RESPONSE (4) NO
- Dynamic Design Criteria
- Stiffness Evaluation
- Expected Earthquake Waves
- Damping
- Damage Control
- Safety Judgement

YES

EXAMINATION BOARD in J.B.C. (5)
- Examination Board

MINISTRY APPROVAL of construction (6)
- Ministry of Construction
- Special Permission

Figure 3:   Seismic Design Procedure of Tall Buildings in Japan.

Seismic Design Procedure

The seismic design procedure for tall buildings is shown in detail in Figure 4. At first, the horizontal earthquake force and its distribution are assumed by AIJ standard as mentioned above (1), and the structural design (3) is determined both by static calculation (2) and test (4). Using observed records of strong earthquakes (6), dynamic response analysis (7) is then conducted with the aid of a computer.

The stress and deformation (8) of the structure are checked and judged (9), and if the results on stress and strain differ from the design criteria or assumption set up (5), they will be fed back into the structural design (3). Final design of the building is obtained after several repetitions of the procedure (10).

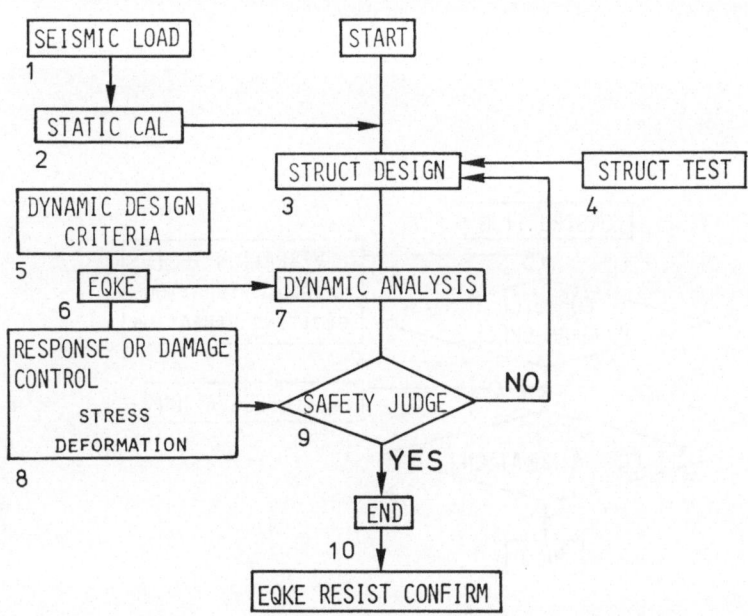

Figure 4: Seismic Analysis of Tall Buildings

Post Construction Study

A post-construction study to assure the adequacy
of the analysis method and the safety of the structure
must be performed even after completion (1) of the
building as shown in Figure 5. In this study, a vibration observation and a vibration test with or without
exciters are carried out (2) . If it is possible to obtain earthquake records by strong motion accelerographs
installed in the building (3), post-earthquake simulation (4, 5) is consequently achieved.

It is most important to compare the dynamic behaviour of the actual building with presumed calculation. If the building vibration can not be explained
satisfactorily by the computer programs, they must have
some defects and need improvement. It is sure, however,
that the seismic design and study procedure themselves
have a large feed-back system, which helps to make -
day by day - remarkable progress in earthquake resistant design.

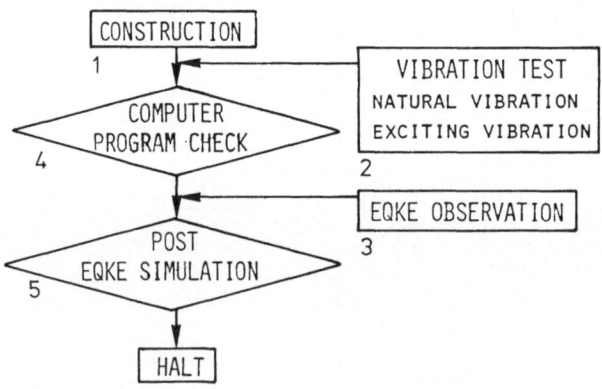

Figure 5:   Study of Post Construction Behaviour of Tall
            Buildings

SEISMIC DESIGN BY MUTO INSTITUTE

Seismic Design Criteria

    Figure 6 shows the seismic design criteria set up by the Muto Institute. The author assumes three classes of earthquake intensity, and against each class, some criteria are established.

    For a minor or moderate earthquake, Class I, which occurs often, the vibration must be controlled to be as small as possible without human disturbance or discomfort.

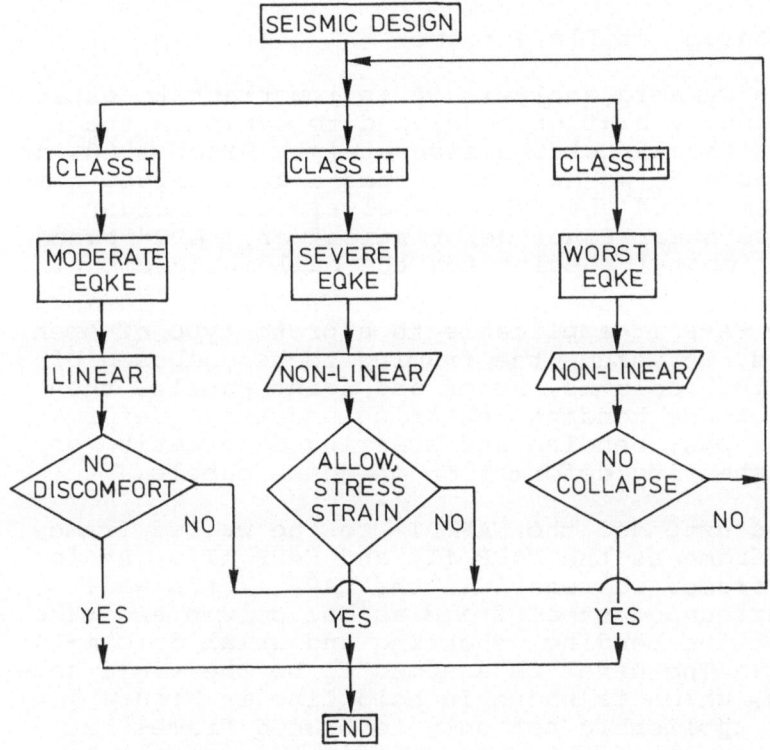

    NOTE: Consideration should be paid to
           the strong wind effects together
           with earthquake motion.

Figure 6: Seismic Design Criteria set up by the Muto
           Institute.

When a severe earthquake of Class II hits, which corresponds to present regulations for the seismic force, the stress in all members must be less than the allowable values.

During and after the worst earthquake hypothetically assumed as Class II, stress induced by the earthquake is so large that the building suffers, to some extent, damages such as residual deformations. Without those damage evaluations and controls, seismic design, in the true sense of the word, can never be established. We are now capable as described hereunder to face the problems which were unavoidably neglected due to difficulties in calculations.

Computer Program for Plane Frame

For the dynamic analysis it is important to establish a rigorous vibration model and to evaluate the stiffness of the structures accurately. Since 1960, as shown in Figure 7, we have developed a generalized computer program named the "Frame Analysis in consideration of Pure-shear Panel deformation" or "FAPP" based on the experimental results for the building element.

In the FAPP I applicable to a proto-type of open steel frames, we assume the framing to be composed of 3 elements, i.e., columns, beams and joint panels, then take into account bending, shearing and axial deformation on columns, bending and shearing deformation on beams, and shearing deformation on joint panels.

We have extended the FAPP I to the walled frame and braced frame as the FAPP II, and FAPP III. As for the walled frame, we regard in the FAPP II the wall with its surrounded steel frame as one column and take into account the bending, shearing and axial deformation on it. The panel is assumed to be the rigid zone of the beam, which is shown in bold line in Figure 7. FAPP III is applicable not only to braced frames but also walled frames after some modification on the beam above the wall.

Computer Program for Three-dimensional Frame

FAPP IV was further developed in order to apply three dimensional framing.

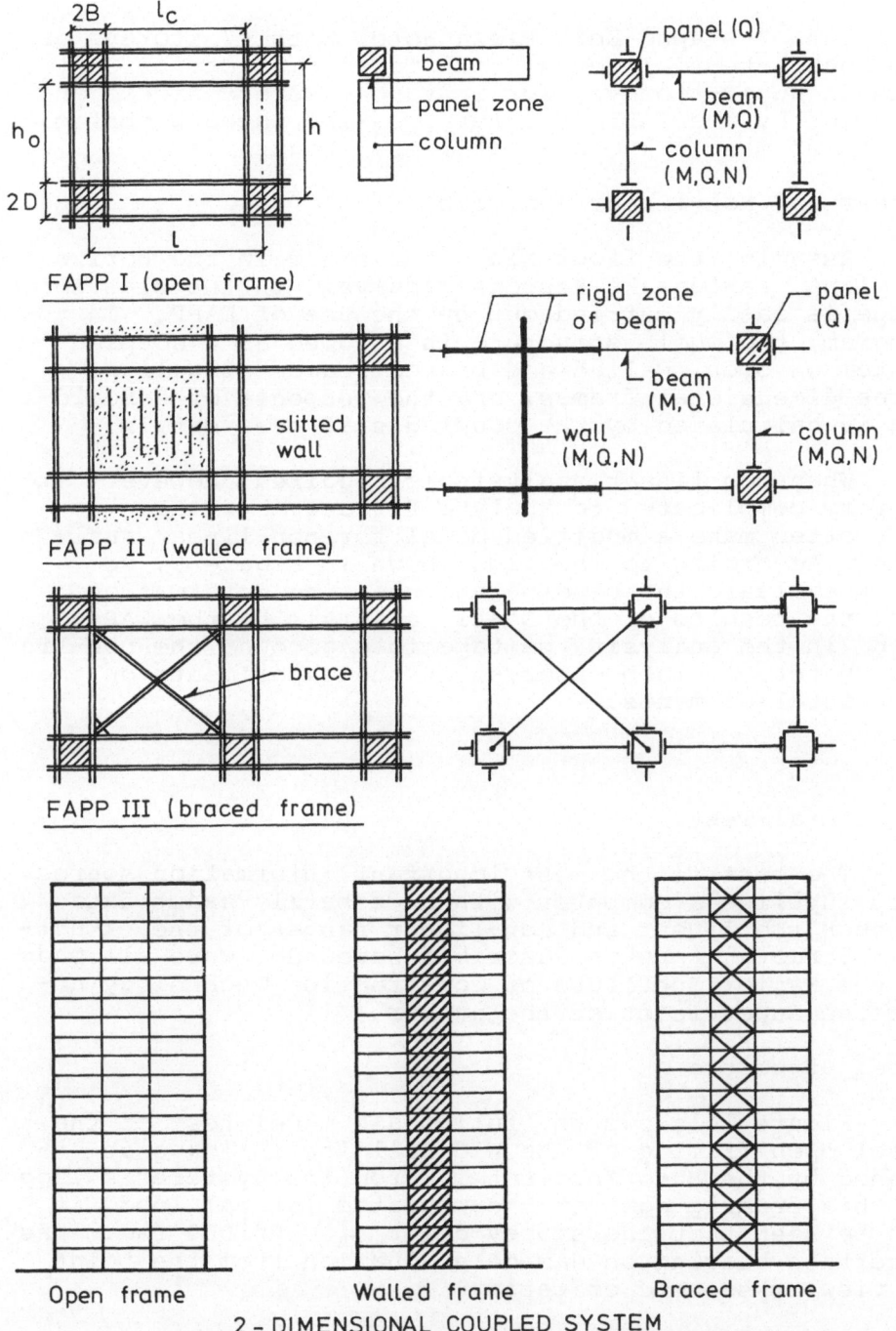

FAPP I (open frame)

FAPP II (walled frame)

FAPP III (braced frame)

Open frame   Walled frame   Braced frame

2 - DIMENSIONAL COUPLED SYSTEM

Figure 7: Development of the Computer Programme "FAPP" (Frame Analysis in consideration of Pure -shear Panel deformation).

The Z-shaped Keio Plaza Hotel with 47 storeys or the Double-tubed International Telecommunication Centre with 32 super-storeys, for instance, were designed and analysed by the FAPP IV after breaking through obstacles due to so many unknown quantities, (Figure 8).

Linear and Non-linear Analysis

Assuming the floor slabs are rigid in the horizontal plane, earthquake response analysis in the elastic range is easily carried out by the use of FAPP. In the program, the whole structure is assumed as a coupled system of open, walled and braced frames as well as three dimensional frames, and the response of a building is calculated by the coupled stiffness matrix.

When non-linear analysis is required, however, it is very complicated to analyze directly by FAPP. We very often make a modified model for non-linear analysis. According to the flow shown in Figure 9, we first evaluate the bending and shearing deformation from the results of the static analysis by the FAPP. Then, in the analysis, we take into account the non-linear hysteresis loop observed in the experiments on the structural elements.

Structural Test

The test is the most important information source that supplies a computer with necessarily exact data through the linear and non-linear ranges of the structure. Structural tests described hereunder were all made by the Kajima Institute of Construction Technology under the supervision of the author.

Steel Structure

Figure 10 shows the full-scale model test of the steel open framing of the WTC Building in Tokyo designed by the Muto Institute. From the hysteresis loop of this framing against the repeated lateral load, it can be seen that the storey drift of 10/1000 rad. has a certain bearing on damage evaluation from the point of view of seismic criteria.

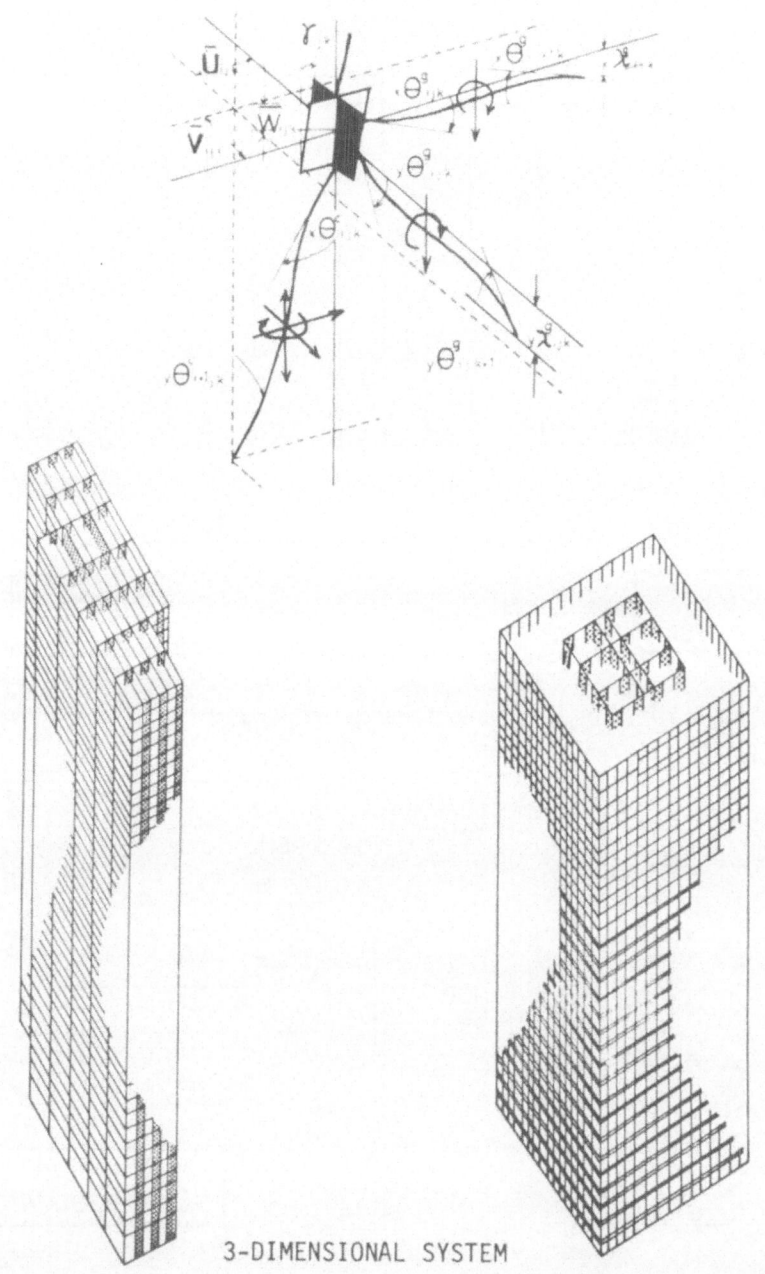

3-DIMENSIONAL SYSTEM

Figure 8: The 3-dimensional Computer Programme FAPP IV

216

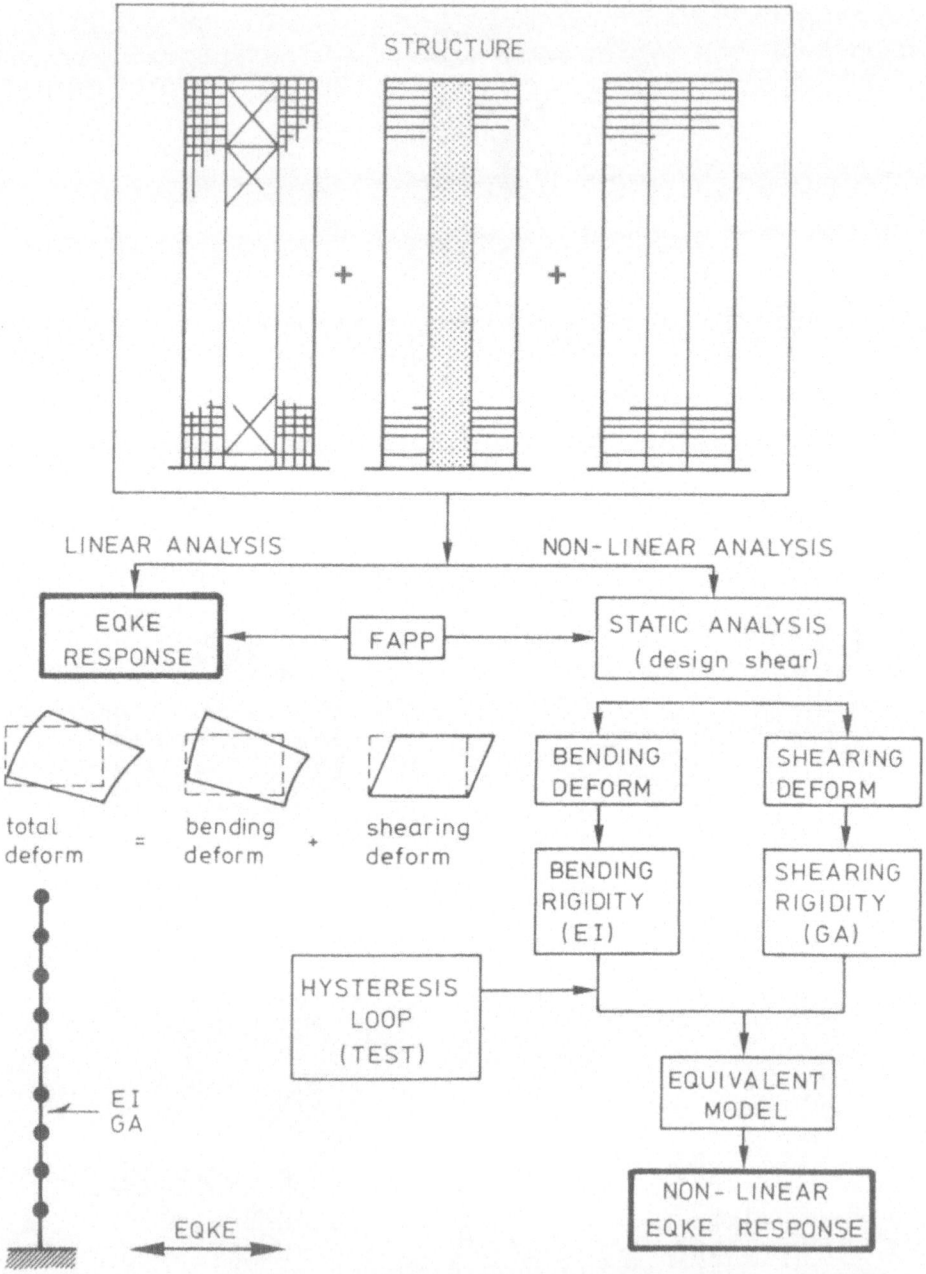

Figure 9: Flow Diagram for Computer Analysis of Linear
and Non-Linear Structures.

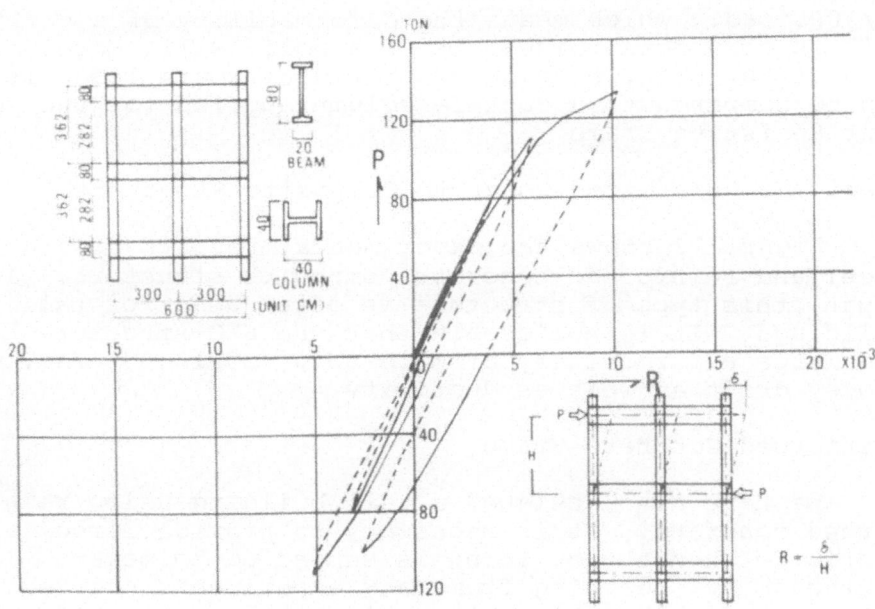

STEEL FRAME

Figure 10:   Full-Scale Model Test of Open Steel Frames
in the Tokyo WTC Building.

Slitted Wall

Slitted walls have been developed by the Muto Institute as ductile resisting shear walls, and used for
many steel structures, such as the 36 storey Kasumigaseki building., the 40 storey WTC building and the 47
storey Keio Plaza Hotel. By slitting the reinforced
concrete wall, it is possible to reduce its initial rigidity and to distribute the burden of resisting any
horizontal forces more favourably between wall and
steel frame. Furthermore, even when the wall is in the
plastic range, ductility is proved to maintain the resisting capacity and capability to withstand a large deformation of the slitted wall.

Figure 11 shows the structural test of the slitted
wall installed into the steel framing of the Keio Plaza
Hotel. On the test specimen of the slitted wall, it is
evident that the cracks are distributed finely with no
major diagonal cracks fatal to an ordinary monolithic
wall. (From the hysterisis loop of slitted wall, it is
recognized that the deformation angle can reach up to
10/1000 rad., which means the deformability of a slitted
wall is more than twice that of a solid wall. Moreover,
without any brittle failures at the ultimate load, it
can be compared to a ductile column applied for the "Moment Resisting Frame",

Steel and Reinforced Concrete Composite Structure

Figure 12 shows the experimental results of a
steel and reinforced concrete composite structure. In
Japan, this type of structure is often used for tall
buildings. It is noticeable that the SRC-structure has
the large deformability of more than 10/10,000 rad. of
storey drift as well as ductility.

Reinforced Concrete Frame

We have also designed a tall building using reinforced concrete. It is necessary to provide large
ductility when the building is subjected to severe
earthquakes. Learning from many experiments carried
out since Tokachi-Oki in 1968, it has been found that
the method of reinforcement has to be improved by using
ties or spirals in addition to conventional hoops for
the column.

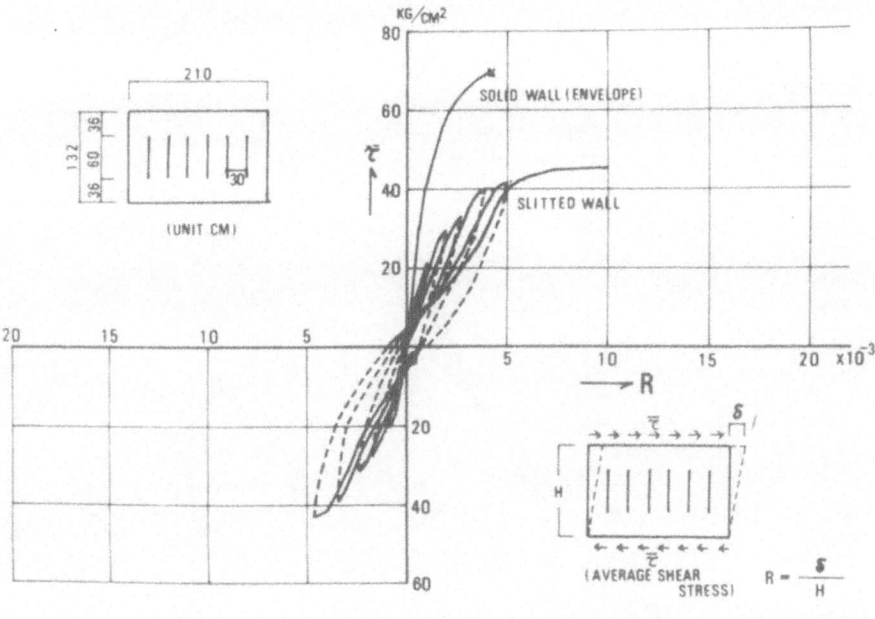

SLITTED WALL

Figure 11:   Structural Test of Slitted Walls Used in
the Keio Plaza Hotel.

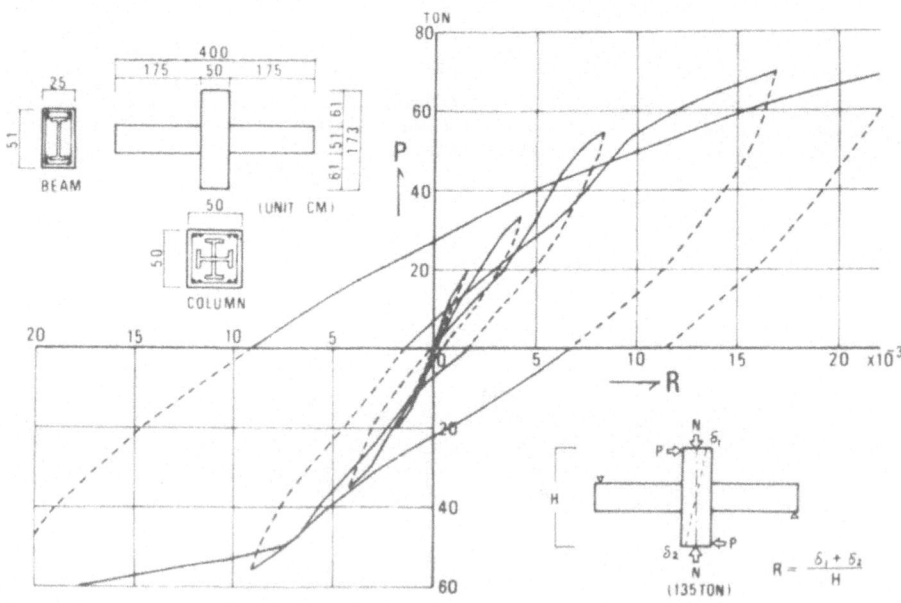

STEEL REINFORCED COMPOSITE COLUMN

Figure 12: Structural Test of RC Composite Column-Beam
Assembly.

Figure 13 shows the structural experiment for a 20 storey reinforced concrete building. Based on experimental findings, the column is designed unprecedentedly to adopt spirals together with hoops as shear reinforcement. Using $1/\sqrt{2}$ scale specimens, the tests have been performed under the alternate low cycle loading. It is recognized that the hysteresis loop is very stable with storey deflection of 10/1000 rad. and that few residual deformations can appear. Ductility and deformability of 50/1000 rad. are satisfactory.

Idealization of Test Results

These test results are idealized to be used for computer analysis. Figure 14 shows the idealized shear rigidity of a steel frame and slitted wall. Assuming large deformation induced by severe earthquake, both are modified to have bi-linear typed non-linearity. The relative resisting behaviour of each element should be noticed.

Figure 15 shows the idealized loop for the current analysis of a 20 storey reinforced concrete structure, where we have assumed a degrading tri-linear property. In the figure, the point (A) indicates the first tensile crack. Up to point (A) the structure is completely elastic. In the region between point (A) and (B) the curve tends to return to the origin (0) when the load is released. Point (B) means that yield has occurred in the beam. Over the point (B), deformation may flow to point (M). If the load decreases at point (M), the rigidity is the same as (OB) until the load becomes zero, and after that, the curve terminates towards the maximum deformation ever experienced.

Soil and Damping Consideration

Figure 16 shows the analytical condition of soil and foundations. The foundation is assumed to be fixed at the base in case of rock and rigid sub-soil condition or for preliminary design analysis, (Figure 16a). Building-soil interaction, however, should be taken into account especially in case of soft sub-soil condition or for the final confirmation as shown in Figure 16b. Mass effect and damping characteristics due to sub-soil are both to be considered, especially with regard to the energy dissipation.

222

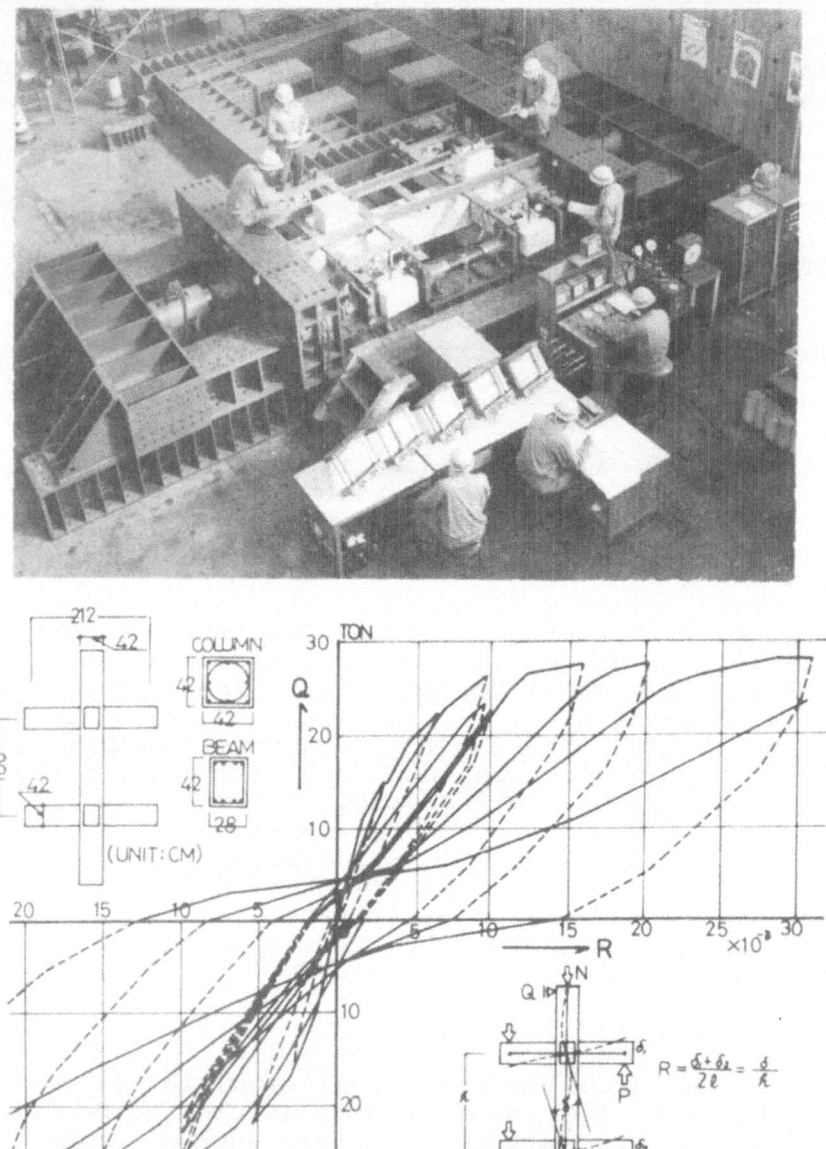

REINFORCED CONCRETE FRAME

Figure 13: Structural Tests of Column Elements in a
20 Storey RC Frame.

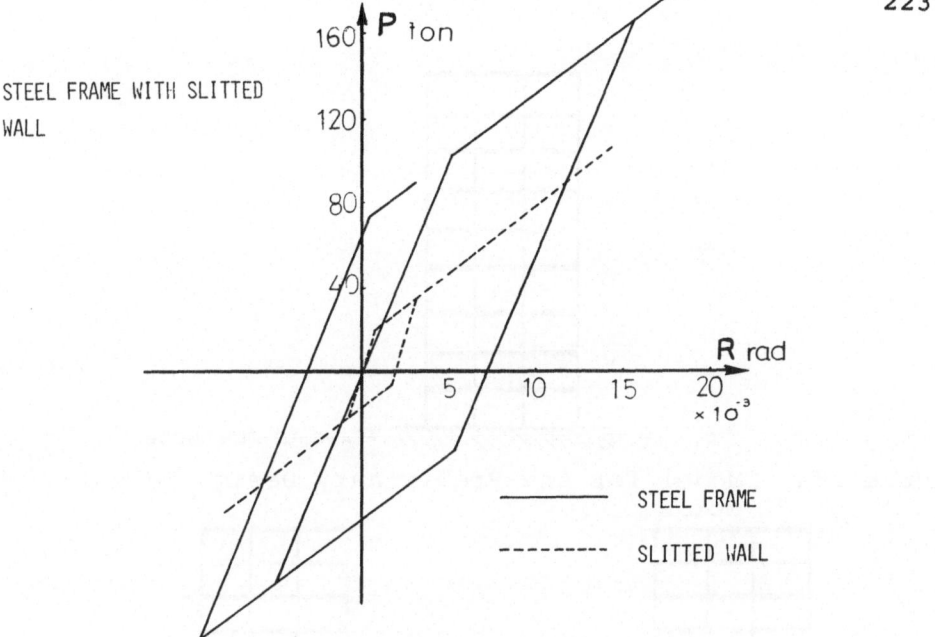

Figure 14: Force-Deformation Idealization for Steel Frames and Slitted Walls.

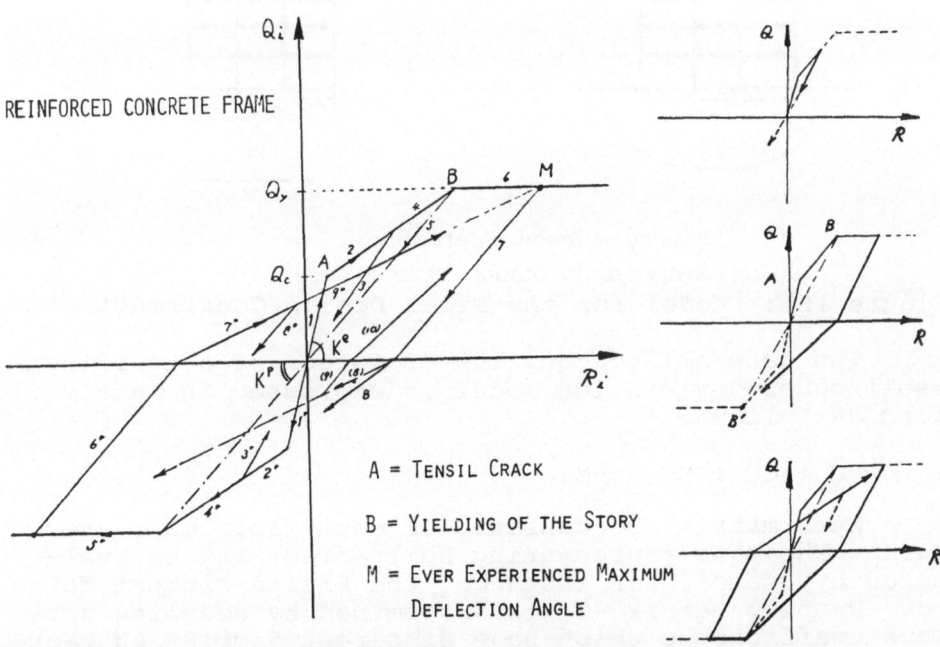

A = Tensil Crack

B = Yielding of the Story

M = Ever Experienced Maximum Deflection Angle

Figure 15: Force-Deformation Idealization for Reinforced Concrete Frames.

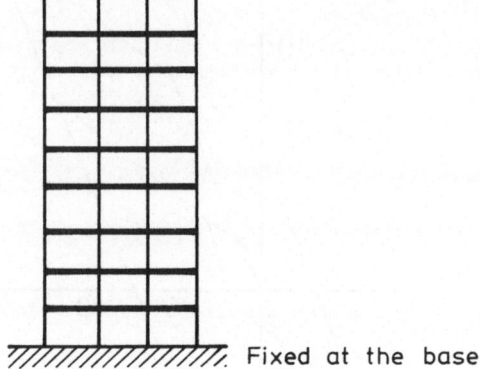

Fixed at the base

Figure 16a: Model for the Preliminary Design Calculation

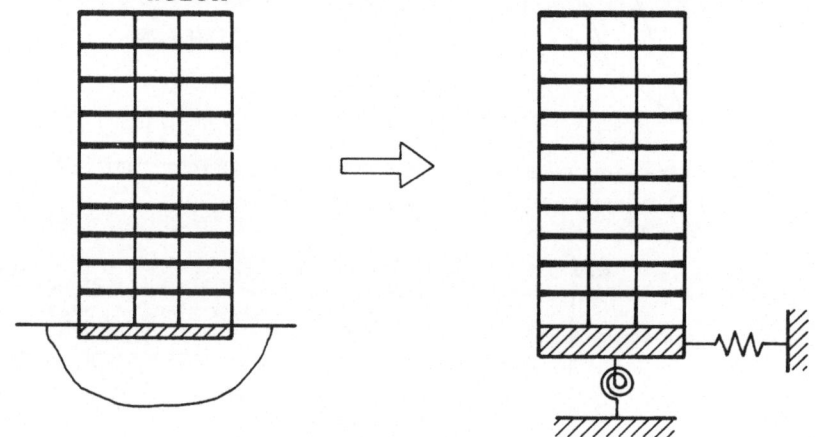

Building-subsoil interaction
(Sway and rotation of the base)
Figure 16b: Model for the Final Design Confirmation

The generalized equation of motion of a building
-soil coupled vibration model is expressed in matrix
form as follows:

$$M\ddot{U} + C\dot{U} + KU = -M\alpha$$

Mass matrix M consists of each floor mass and
soil. The mass representing soil effect may be evaluated by use of, for instance, the Finite Element Method. Damping matrix C is determined by coupling various coefficients which have different damping characteristics one by one. Stiffness matrix K also includes soil properties resulting in sway or rotation of
the foundation of the building.

Examples of Post Earthquake Simulations

On July 1., 1968, shortly after the completion of
the 36 storey Kasumigaseki building, the North-Tokyo
earthquake struck.   Excited motions of the building
were measured by the SMAC (Strong Motion Acceleration
Seismograph) installed on several floors. Against input ground motion with maximum acceleration of 25 gal ,
maximum accelerations of the upper floors were distributed variously, and the natural periods for the 1st,
2nd and 3rd modes proved to be 4.0 sec., 1.2 sec. and
0.6 sec. The computer-calculated periods showed very
close agreement. Values for the 1st, 2nd and 3rd modes
in the low stress states were 4.02 sec., 1.14 sec. and
0.55 sec. Furthermore, with the same program, a response analysis of the building was performed using the
input earthquake waves recorded at the foundation floor,
and resulted in good agreements between the calculation
and observation as shown in Figure 17.

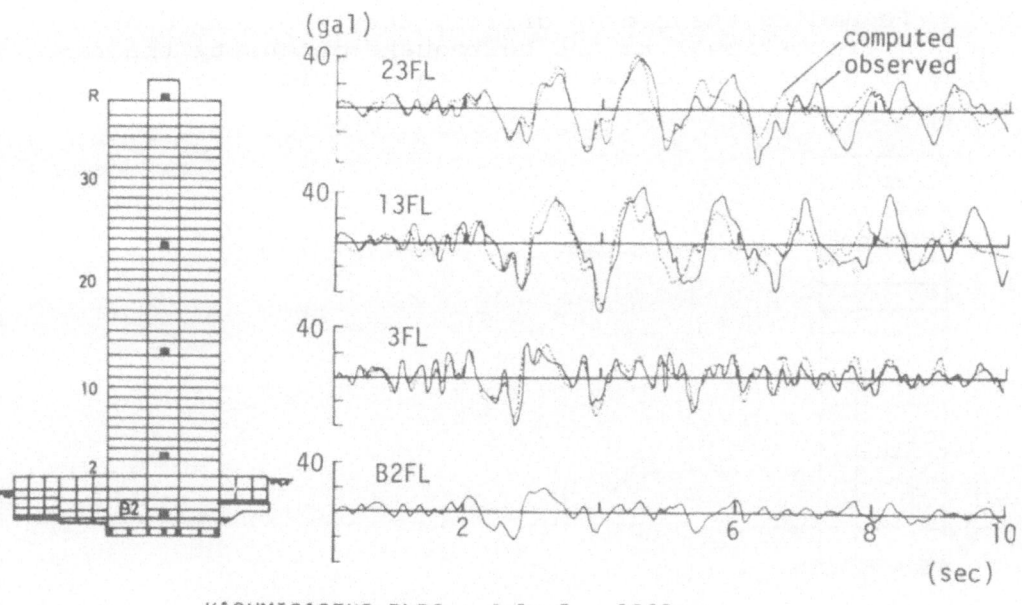

KASUMIGASEKI BLDG., July 1 , 1968
NORTH-KANTO EARTHQUAKE

Figure 17:   Observed and Computed Earthquake Response
             of the Kasumigaseki Building in Tokyo.

As for the earthquake simulation with larger ground acceleration, the 16 storey K.I.I. building is the newest example. That is, the building suffered a severe excitation due to the San Fernando earthquake in Los Angeles on February 9, 1971. Ground acceleration with the maximum 133 gal in transverse direction of the building was amplified up to 166 gal at the 6th floor and 172 gal at the roof. Through all the analytical procedures, again better coincidence between the model and the measured results were obtained as shown in Figure 18.

Examples of Response Controls

We have undertaken to design and analyze many high-rise buildings in the way described above. Examples of maximum response values of four tall buildings against the El Centro and Taft earthquakes are listed in Table 1. All of these values are appropriate for the design criteria defined by classification as I, II and III .

Regarding the storey drifts, they are - more or less - proportional to the earthquake intensity the ratios of which are 1:2:4 . The same ratios due to

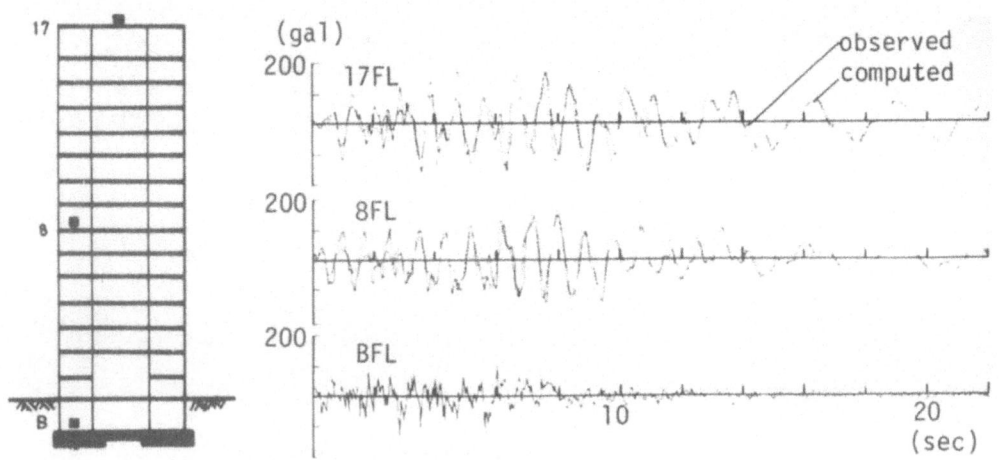

K.I.I. BLDG., February 9, 1971
SAN FERNANDO EARTHQUAKE

Figure 18: Observed and Computed Earthquake Response of the K.I.I. Building in Los Angeles.

Building	Earth-quake	Class*	Max Storey Drift** (cm)		Max Acc. (gal)	
			Total	Slitted Wall ***	Top	Middle
Kasumigaseki Building	EL CENTRO 1940 (NS)	I	0.7	0.2	100	60
		II	1.3	0.5	130	90
		III	2.4	1.0	240	180
Height-147m Storeys-36	TAFT 1952 (EW)	I	0.4	0.2	80	60
		II	1.0	0.5	130	120
		III	1.9	1.1	260	220
Tokyo World Trade Center Building	EL CENTRO	I	0.7	0.3	130	70
		II	1.3	0.7	160	90
		III	2.3	1.5	240	190
Height-152m Storeys-40	TAFT	I	0.4	0.2	80	70
		II	0.8	0.7	140	120
		III	1.6	1.5	280	240
The Keio Plaza Hotel	EL CENTRO	I	0.6	0.2	100	60
		II	1.2	0.5	170	100
		III	2.5	1.0	340	210
Height-170m Storeys-47	TAFT	I	0.5	0.2	90	70
		II	1.0	0.4	170	140
		III	1.9	0.9	350	290
Shinjuku Mitsui Building	EL CENTRO	I	0.4	0.1	60	40
		II	0.7	0.2	110	80
		III	1.4	0.5	220	160
Height-212m Storeys-55	TAFT	I	0.4	0.2	60	50
		II	0.7	0.3	120	90
		III	1.4	0.6	240	180

NOTE: * I-100 gal, II-200 gal, III-400 gal.
 ** on Typical Storey Height.
 *** Storey Drift of Slitted Wall due to Shear Deformation.

Table 1: Earthquake Response Characteristics of the 4 Tallest Buildings in Japan.

shear deformations of slitted walls are somewhat higher. Maximum response accelerations are considerably smaller than those input values as a special feature of the tall building. When the El Centro earthquake is applied to the Shinjuku Mitsui Building, for instance, the response accelerations at the top floor become only 60, 110, and 220 gal respectively. Those accelerations at the middle floor, moreover, become much smaller

EXAMPLE OF DESIGN OF TALL BUILDING

The Shinjuku Mitsui Building

The construction site of the Shinjuku Mitsui building (S.M.B.) is located at the new business centre of Shinjuku in Tokyo. The Keio Plaza Hotel of 47 storeys has already been completed there, and some other high-rise buildings including the International Telecommunication Center of 32 super-storeys are under construction, (Figure 19).

Figure 19:   The 55 Storey Shinjuku Mitsui Building

S.M.B. is a 55 storey, 225 metre high office
building which will be the tallest in Japan. Typical
floor areas are 58.4 x 44.5 square metres. The construction work was started in April 1973, and will be
completed in September 1974.

Focal Point of the Design

In designing S.M.B., minimization of shaking
during strong winds becomes no less an important problem than safety against strong earthquakes. Because
many Japanese are very nervous about tremors of high
-rise buildings from typhoons which frequently occur,
the author made it a basic design concept to give the
building sufficient stiffness as well as strength and
ductility.

In order to fulfil these demands, the author, at
the beginning, aimed to design the stiffness of the
building to make the fundamental natural period less
than $T_1 = 0.1 \times N$ (N: the number of the storey), that
is, 5.5 seconds. As a result of the calculations at
the preliminary stage, it was found that the fundamental natural period in the transverse direction was not
sufficient. We were then obliged to study various
structural systems, to reach the solution regarding the
stiffness.

Design and Study Procedures

Scheme Proposed by Architect. The structural
system in the typical framing plan proposed by the
architect is shown in Figure 20. The columns are mostly H-shaped and the girders are pin-jointed to them.
Braces are arranged at both ends as symbolic elements
in the architectural design. Columns and girders are
of H-shaped steel having 400 mm flanges and 800 mm
depth. It is found that the stiffness is too small
(T_1 = 6.30 sec.) and that the braced frame is excess-
ively loaded (37.5%) with the lateral forces.

One of Revised Studies. Figure 21 shows one of the
revised structural systems, where the direction of inner H-columns is turned 90 degrees and the girders
are rigidly jointed to the columns. Here it is found
that the stresses are excessive though the stiffness is
a little improved, (T = 5.26 sec.), and that the average steel quantity of the tower portion is 120 kg/m^2.

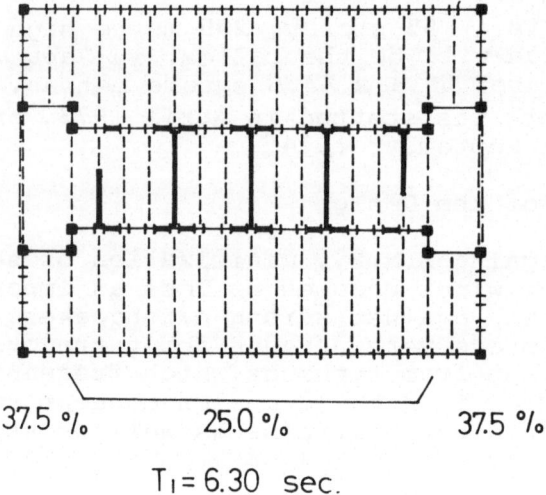

37.5 %   25.0 %   37.5 %

$T_1 = 6.30$ sec.

Figure 20: Original Structural System Proposed for the
SM-Building

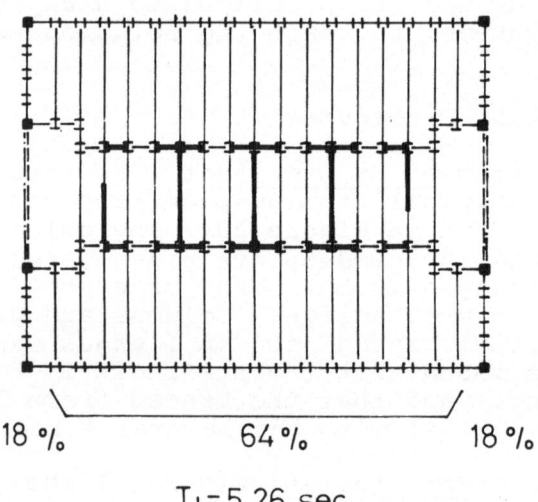

18 %   64 %   18 %

$T_1 = 5.26$ sec.

Figure 21: One of Revised Structural Systems for the
SM-Building

Final Design. After these studies, we finally obtained the structural system as shown in Figure 22. Almost all of the columns are replaced by welded box-type
ones (500 x 500 mm), and the girders are rigidly
jointed to the columns. The main transverse framing
structure consists of three kinds of framing: walled
frames, braced frames and other open frames.

Owing to these final design results, we were able
to realize that the stiffness is quite satisfactory
(T_1 = 5.08 sec.), and the lateral forces are properly
distributed to all framings. Stresses of all the members are, of course, within allowable limits provided
that the average steel quantity of the tower portion is
115 kg/m^2.

Final Design Analysis of the Structure

The design analysis was carried out for the final
design structure on the static frame analysis against
seismic force and the earthquake response. The results
in the transverse direction are discussed hereunder.

Stress Diagram for Seismic Load. The bending moment, shearing and axial forces induced by the seismic
force defined by the AIJ-standard are shown in Figure
23 of the stress diagram of a section of the braced
frame. Squares located at nodal points represent joining panels.

Stress states of each member induced by the lateral
force are comprehensive. As for the axial force of the
super-bracing member, it amounts to about 300 tons
which means that the burden of this member is properly
20 percent of the total lateral force.

Vibration Mode. The fundamental vibration periods
and modes for the 1st, 2nd and 3rd modes are shown in
Figure 24, where the equivalent shear modulus of a reinforced concrete slitted wall is assumed to be G =
15 ton/cm^2 as in case of larger amplitudes against a
severe earthquake. The fundamental 1st period is 5.08
sec., which is far smaller than the upper limit of 5.5
sec. determined previously as the target.

Linear Response. The results of a linear response
calculation against earthquakes with maximum acceleration of 100 gal in Class I are shown in Figure 25.
Four earthquake records, El Centro 1940 (NS), Taft 1952
(EW), Tokyo 1956 (NS), and Sendai 1962 (NS), are adopted as input.

232

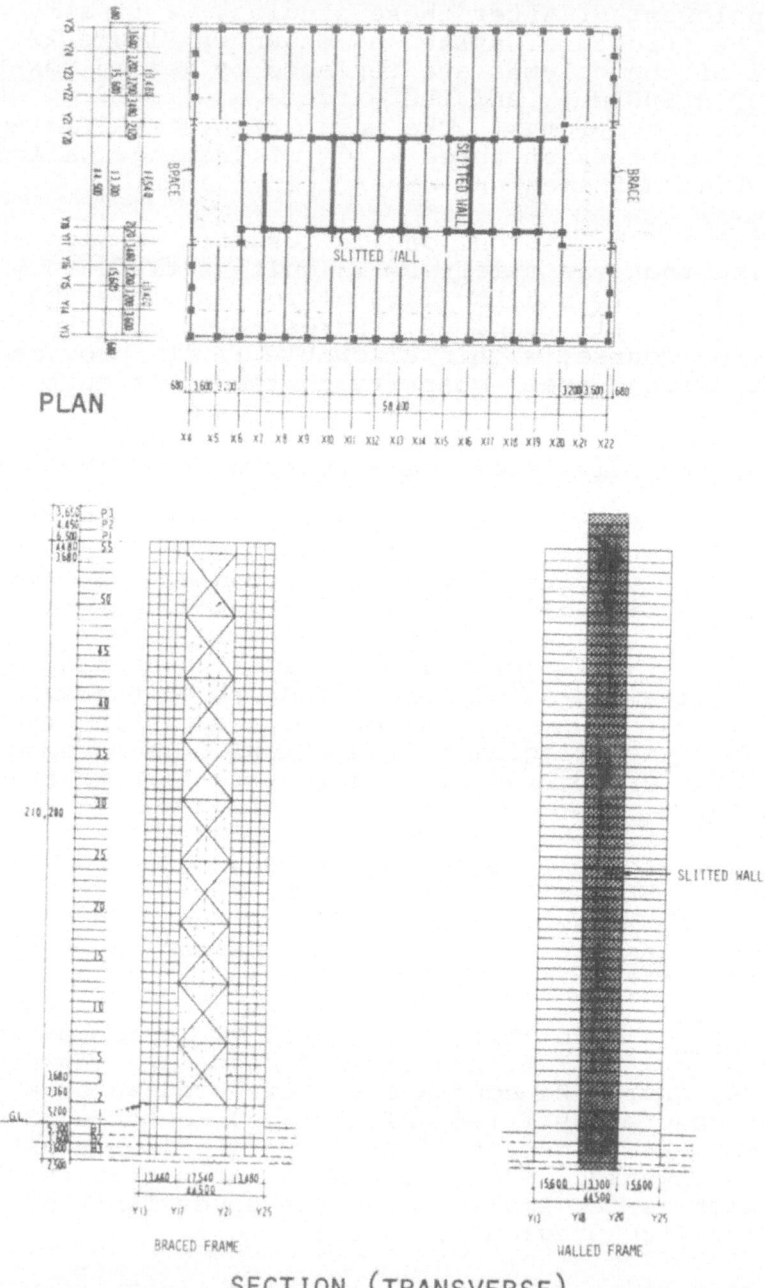

PLAN

SECTION (TRANSVERSE)

Figure 22: Final Structural System Adopted for the SM
-Building

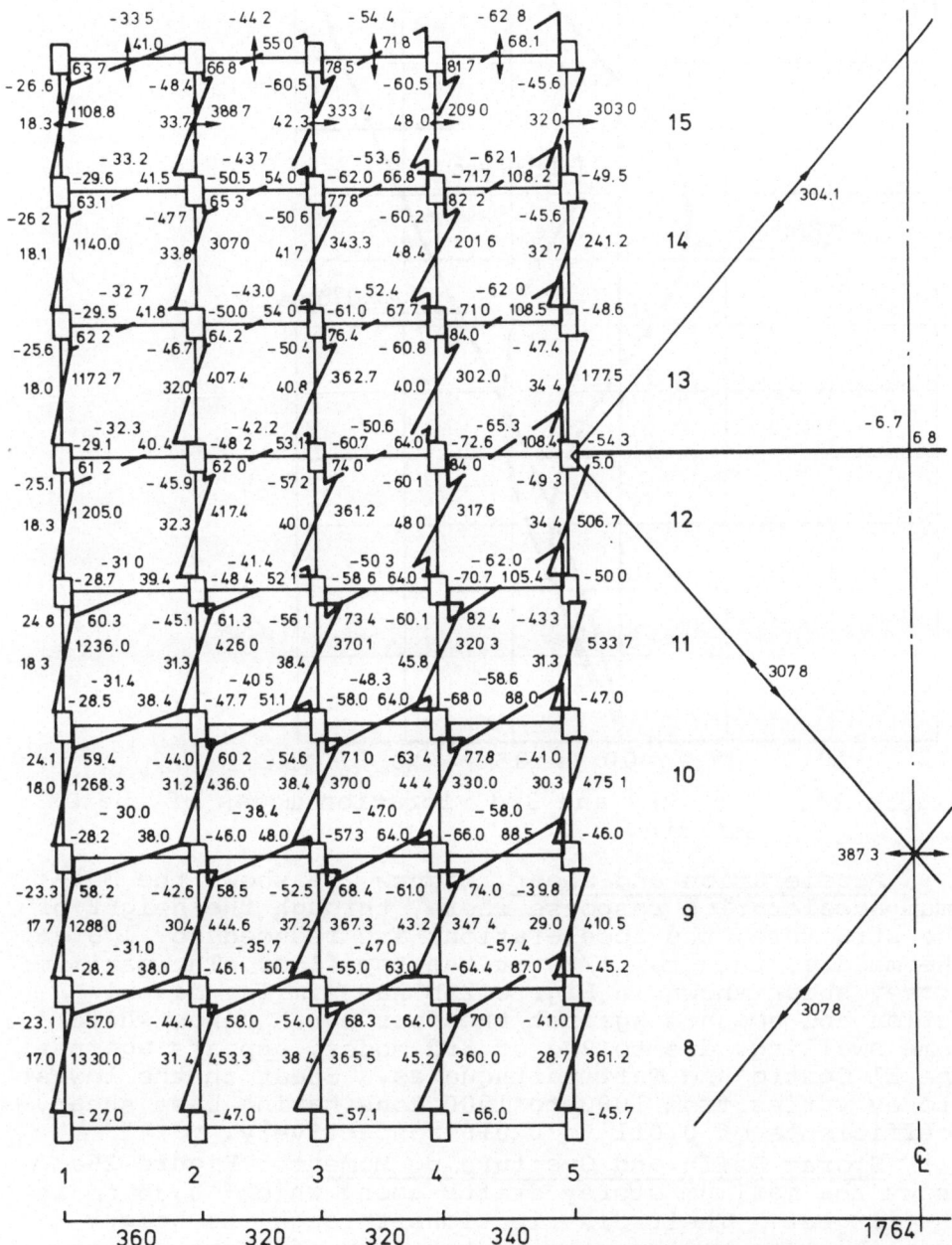

Figure 23: Stress Diagram for the Top 8 Storeys of the SM Building induced by the AIJ Design Seismic Loads.

234

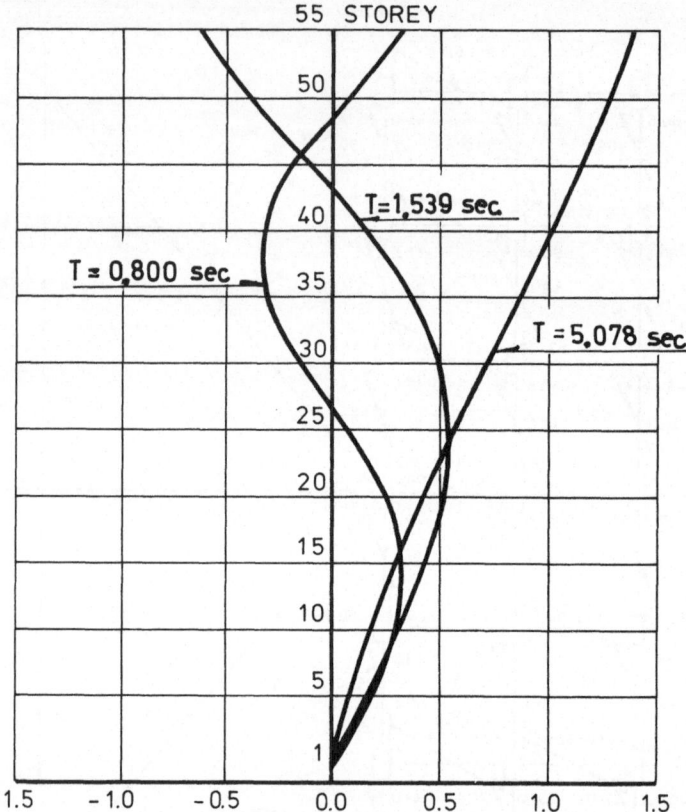

Figure 24:   1st, 2nd and 3rd vibration modes of the SM
             Building.

Acceleration and Shear. Figure 25a shows the maximum acceleration response where, through the height of
the structure, the accelerations are reduced to 1/3 at
the middle, then to 1/2 at the top floor. The maximum
storey shear shown in Figure 25b has similar distribution and amounts against the four earthquakes while
some swelling, due to 2nd or 3rd modes, appears against
the El Centro and Taft earthquakes. Shear in the lowest
storey varies from 1000 to 1500 tons having base shear
coefficients of 0.011 to 0.016 respectively.

Storey Drift and Overturning Moment. Figure 26a
shows the maximum storey drifts among which 0.35 cm is
the largest. Their distributions through the height
are likely to be the same in the form of storey drift
angle although a discontinuity is shown in Figure 26a
due to a higher storey height at the 2nd floor. Overturning moments shown in Figure 26b are gradually and
proportionally decreasing through the height of the
structure while slight swelling due to the higher modes
can appear as for their distributions.

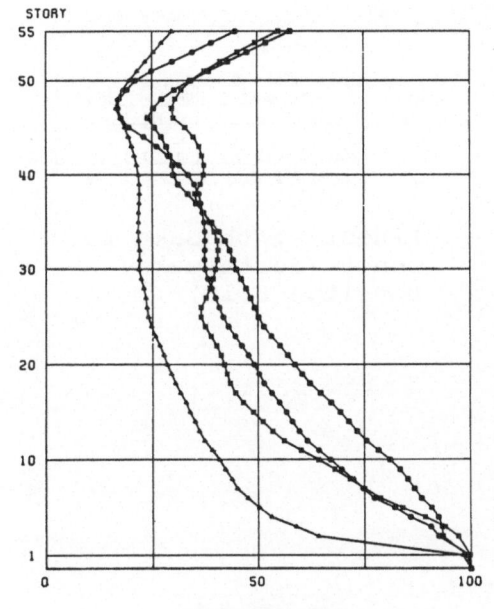

Linear earthquake res-
ponse (Earthquake In-
tensity: 0.1G)

MARK	EARTHQUAKE	DAMPING	αMAX
	EL CENTRO (NS)	0.030	100.
	TAFT (EW)	0.030	100.
	TOKYO 101 (NS)	0.030	100.
	SENDAI 501 (NS)	0.030	100.

Figure 25a: Maximum Acceleration Response of the SM
Building.

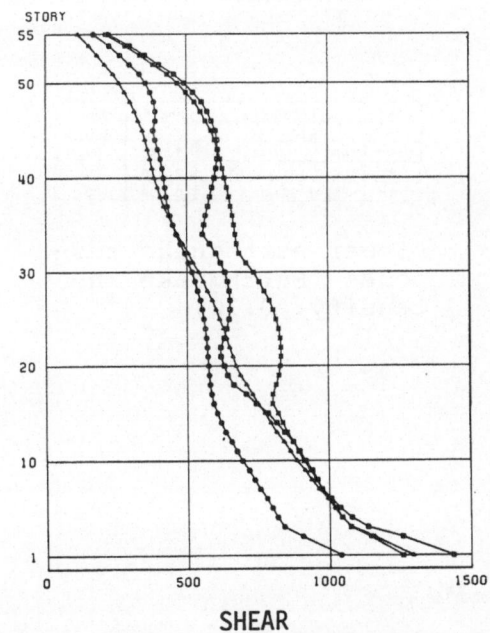

Linear earthquake res-
ponse (Earthquake In-
tensity: 0.1G).

MARK	EARTHQUAKE	DAMPING	αMAX
	EL CENTRO (NS)	0.030	100.
	TAFT (EW)	0.030	100.
	TOKYO 101 (NS)	0.030	100.
	SENDAI 501 (NS)	0.030	100.

Figure 25b: Maximum Storey Shears in the SM Building.

236

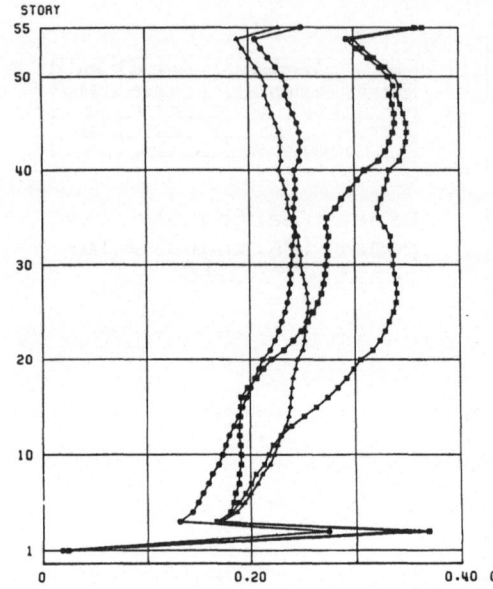

MARK	EARTHQUAKE	DAMPING	αMAX
—•—	EL CENTRO (NS)	0.030	100.
—•—	TAFT (EW)	0.030	100.
—•—	TOKYO 101 (NS)	0.030	100.
—•—	SENDAI 501 (NS)	0.030	100.

Linear earthquake response (Earthquake Intensity: 0.1G)

Figure 26b: Maximum Storey Drifts in the SM Building.

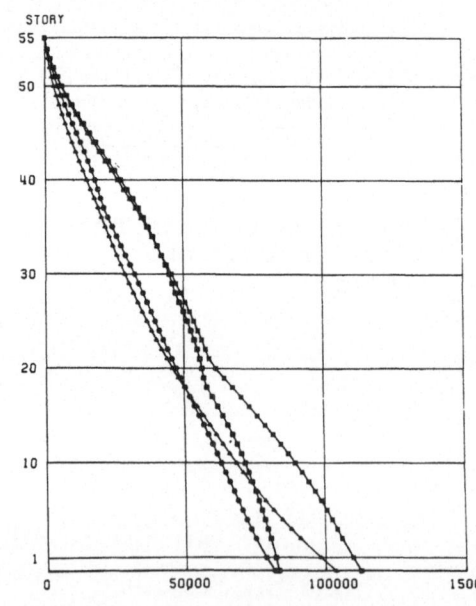

MARK	EARTHQUAKE	DAMPING	αMAX
—•—	EL CENTRO (NS)	0.030	100.
—•—	TAFT (EW)	0.030	100.
—•—	TOKYO 101 (NS)	0.030	100.
—▲—	SENDAI 501 (NS)	0.030	100.

Linear earthquake response (Earthquake Intensity: 0.1G)

Figure 26b: Maximum Overturning Moments in the SM Building.

Non-linear Response. Corresponding to the criteria
on Class II and Class III earthquakes, non-linear response analyses with maximum accelerations of 300 gal
are carried out. The results of the El Centro earthquake are most dangerous, and are shown in Figure 27.
The results on 100 gal acceleration are also dotted for
reference.

Shear and Storey Drift. Maximum response shears
are increased in accordance with the input earthquake
intensities as seen in Figure 27a. Steel frames remain
elastic during a 300 gal earthquake while slitted
walls slightly enter into plastic range at some portions through the height. All the storey shears are
safely within the bearing capacity of the whole structure against a 500 gal earthquake, while steel frames
partially enter into plastic range. Maximum storey
drifts are about 1.0 and 1.8 cm against the earthquake with 300 gal and 500 gal respectively as seen
in Figure 27b, which have no unfavourable effects on
the secondary building elements such as curtain walls,
partitions and so on.

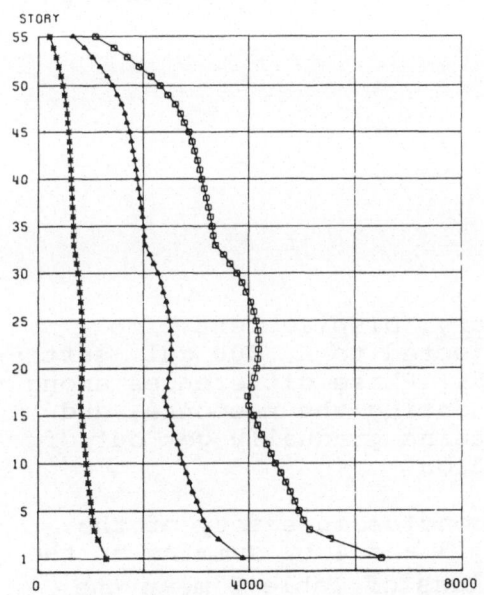

MARK	EARTHQUAKE	DAMPING	q_{MAX}
—*—	EL CENTRO 1940 (NS)	0.030	100.
—▲—	EL CENTRO 1940 (NS)	0.030	300.
—▣—	EL CENTRO 1940 (NS)	0.030	500.

Non-linear earthquake
response (El Centro
Earthquake Intensity:
0.1, 0.3, 0.5G).

Figure 27a: Maximum Shear Response of the SM Building.

238

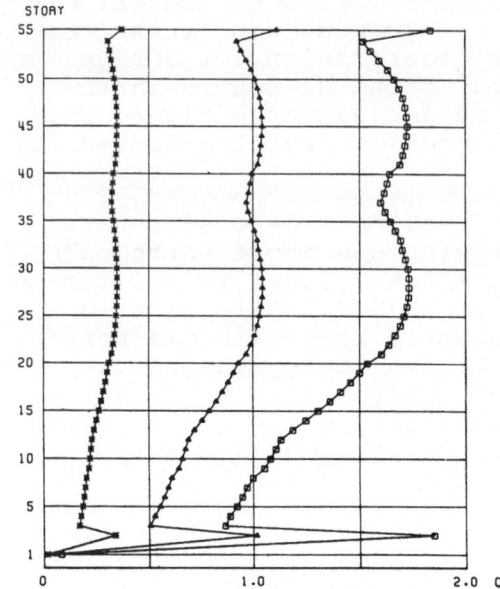

MARK	EARTHQUAKE	DAMPING	qMAX
—■—	EL CENTRO 1940 (NS)	0.030	100.
—▲—	EL CENTRO 1940 (NS)	0.030	300.
—□—	EL CENTRO 1940 (NS)	0.030	500.

Non-linear earthquake
response (El Centro
Earthquake Intensity:
0.1, 0.3, 0.5G).

Figure 27b: Maximum Storey Drifts in the SM Building.

Displacement Time History. Displacement time
histories of each floor subjected to a 300 gal earthquake are shown in Figure 28. Phase differences among
vibrative amplitudes appear during the response, and
sinusoidal displacement patterns gradually get out of
shape through the time duration.

Safety Estimation. In conclusion safety of the
S.M.B. is listed in Table 2, 3 and 4 by summing up the
entire discussions. The values of Table 2 mean the
maximum acceleration when the steel frame possibly
reaches the elastic allowable limits. The values of
Table 3 mean that, with that maximum acceleration of
the earthquake, the storey shear should have been in-

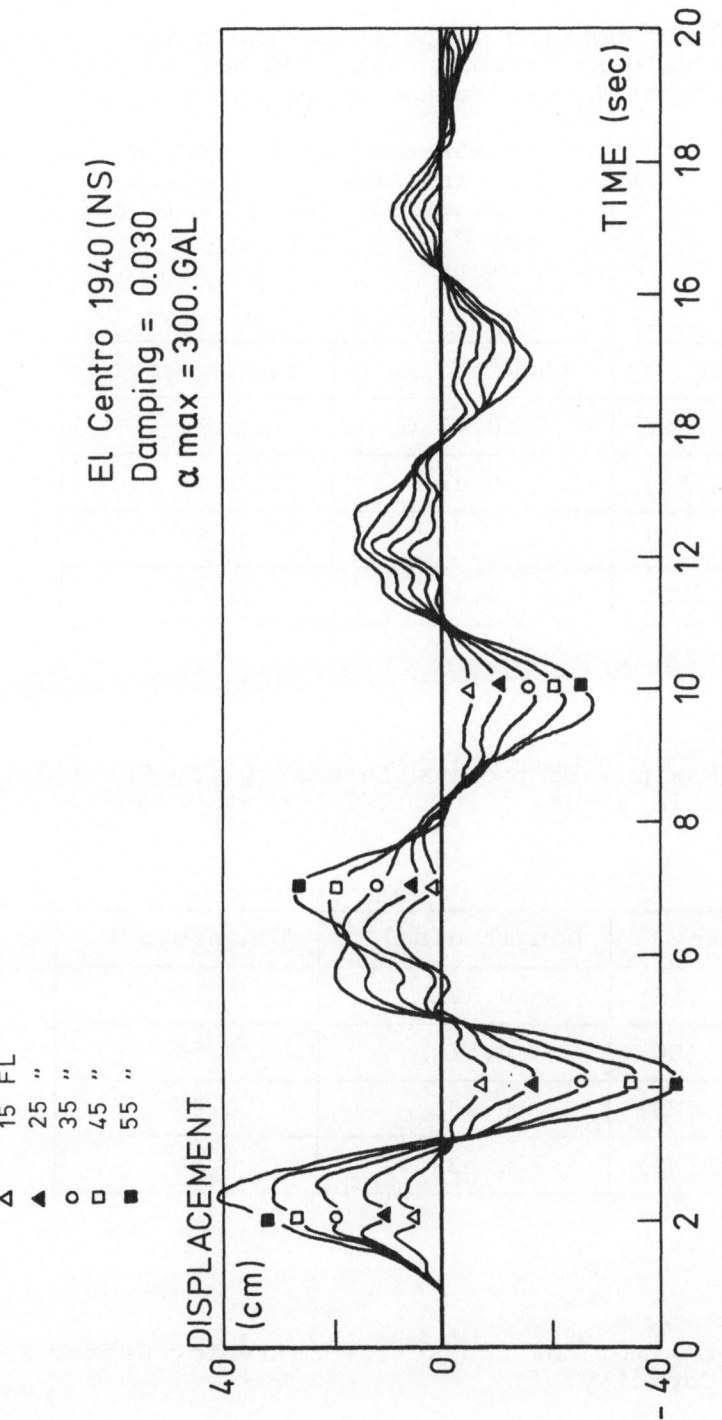

Figure 28: Displacement Time Histories of the 15th, 25th, 35th, 45th and the 55th Floors of the SBI Building.

duced up to the bearing capacity of the building structure. These values are more than 400 gal and 500 gal
which significantly secure the safety of the building.

Against the severe earthquake of 500 gal, maximum
storey drifts are listed in Table 4. Those values are
very small and enough to satisfy the seismic design
criteria.

Earthquake	Longitudinal	Transversal
EL CENTRO (NS)	0.43 G	0.46 G
TAFT (EW)	0.41	0.44
TOKYO 101 (NS)	0.53	0.58
SENDAI 501(NS)	0.43	0.45

Table 2: Maximum Earthquake Intensity for Steel Allowable Stress.

Earthquake	Longitudinal	Transversal
EL CENTRO (NS)	0.52G	0.56G
TAFT (EW)	0.51	0.53
TOKYO 101 (NS)	0.63	0.68
SENDAI 501(NS)	0.53	0.56

Table 3: Maximum Earthquake Intensity for Storey Shear
Capacity.

Earthquake	Longitudinal		Transversal	
	Total	Wall	Total	Wall
EL CENTRO (NS)	1.80	1.10	1.73	0.64
TAFT (EW)	1.68	1.18	1.70	0.70
TOKYO 101 (NS)	1.27	0.88	1.24	0.49
SENDAI 501(NS)	1.32	1.10	1.28	0.64

Table 4: Maximum Storey Drift (for 0.5G) (2nd and 55th Floors are excluded Because of Larger Floor Height than Typical).

Animated Earthquake Response of S.M.B.

A graphic display device, the so-called COM (Computer Output Microfilming) has been utilized in order to express vividly the dynamic behaviour of the structures during earthquakes. We have made so many films from COM that reviews and discussions on the seismic design of the structure would be possible from various points of view.

Among many results on the earthquake response of the final design structure of S.M.B., excitations in transverse direction subjected to the El Centro earthquake with 400 gal maximum acceleration are shown in Figures 29-31. The displacement, storey shear and overturning moment at each storey are presented in order with the lapse of time which is marked upper right in the figures.

Compared to the scale of the building structure, excited displacement of the structure from the fixed axis is enlarged 50 times. Bold lines in the figures of storey shear and overturning moment represent transient values corresponding to the marked time, while dotted lines are memorized maximum values ever experienced.

The building structure itself looks as if it were a hula dancer. And it is also noticeable that all the maximum values induced by the El Centro earthquake happen to occur within 5 seconds of the beginning of the ground motion.

242

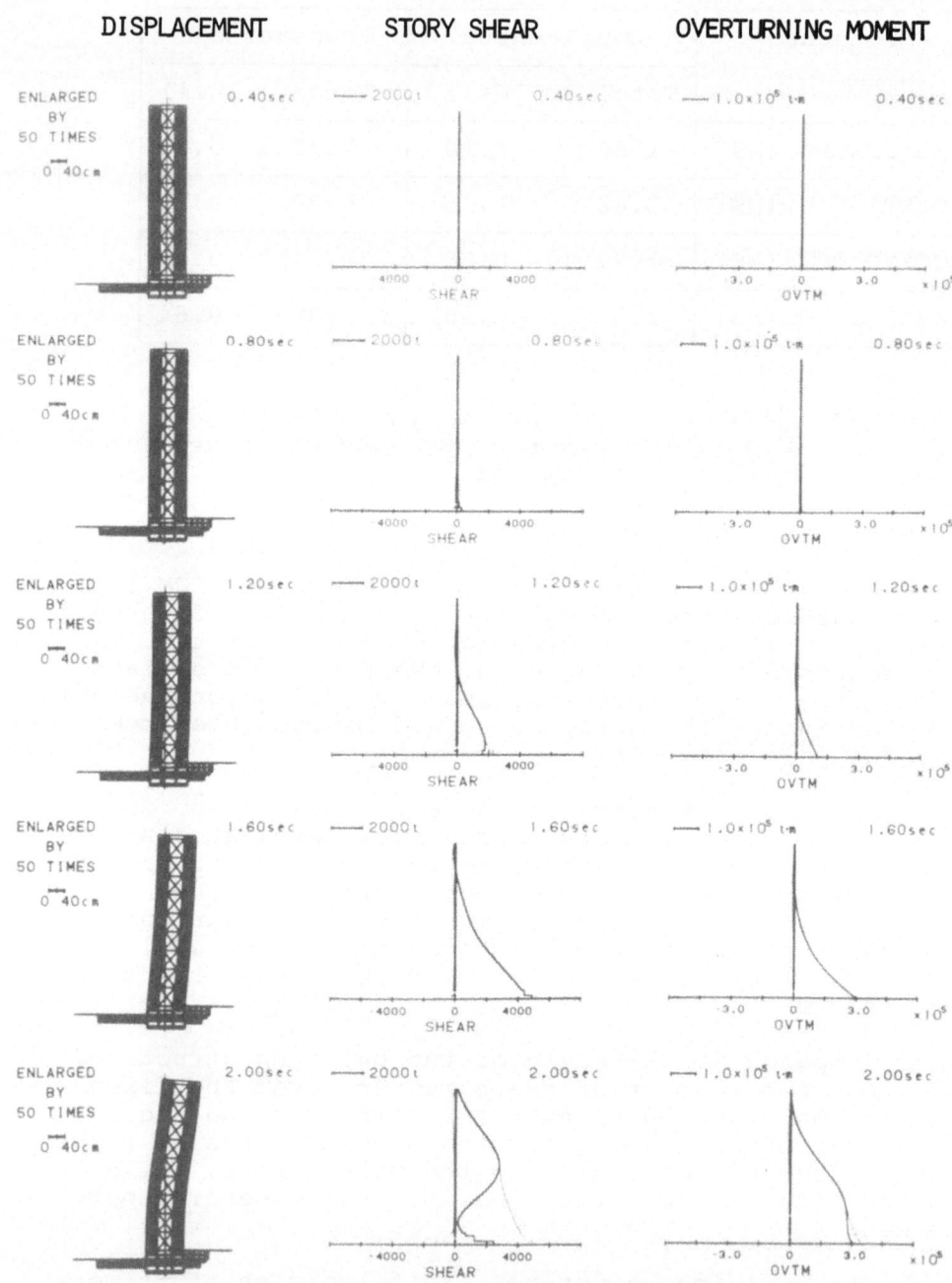

Figure 29: Animated Earthquake Response of the SM
Building (0.4 to 2.0 sec.).

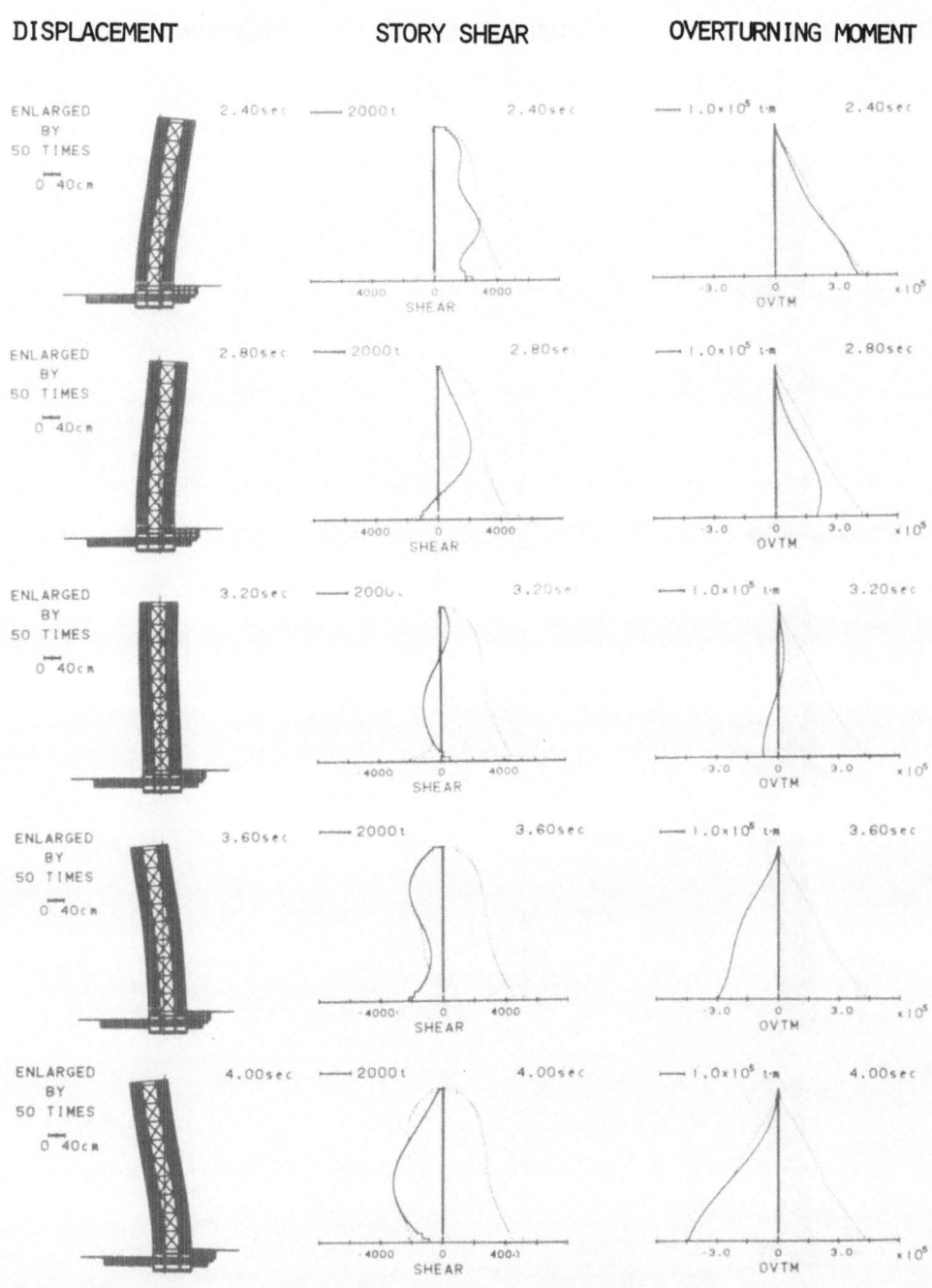

Figure 30: Animated Earthquake Response of the SM Building (2.4 to 4.0 sec).

244

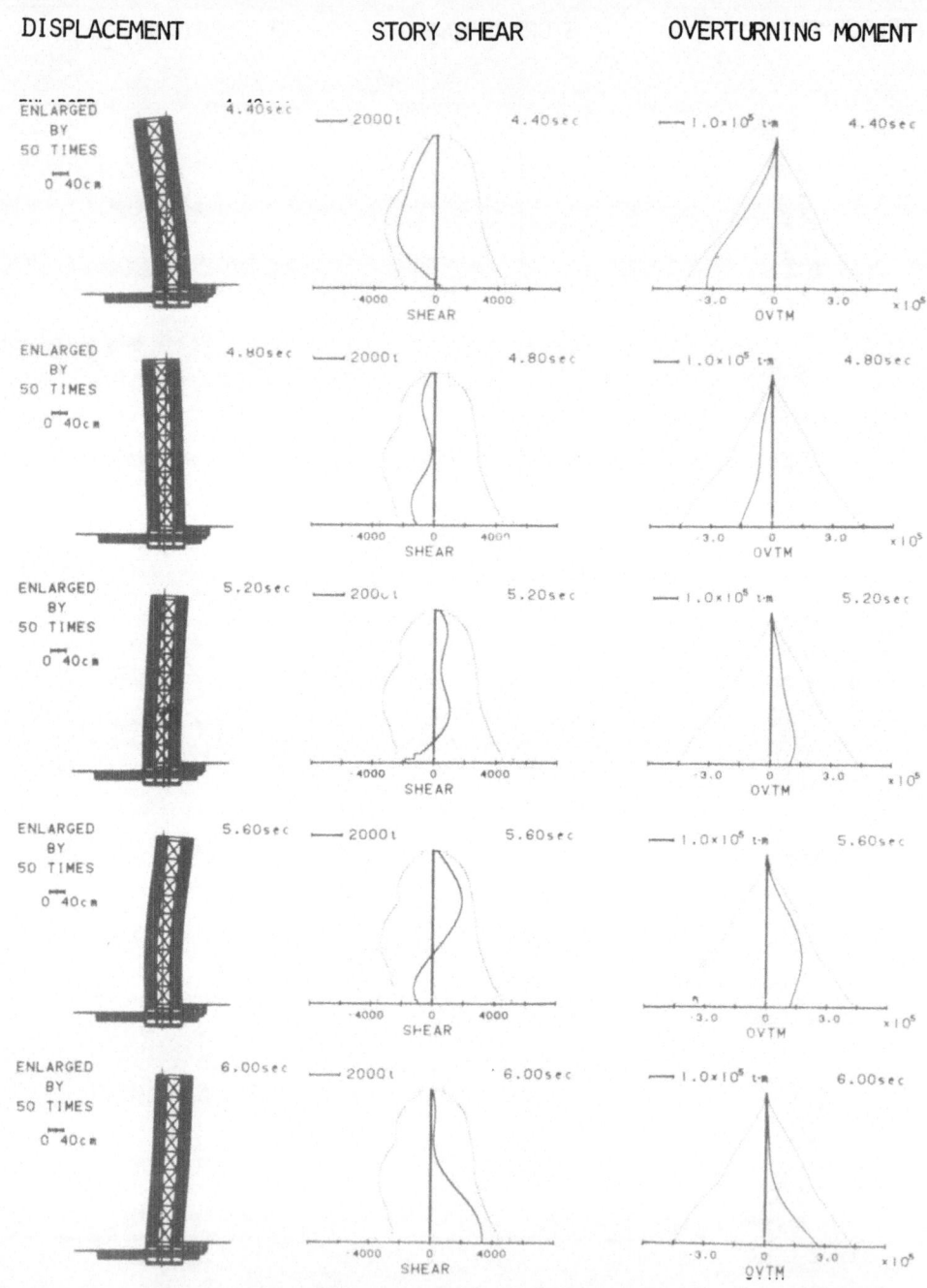

Figure 31: Animated Earthquake Response of the SM
Building (4.4 to 6.0 sec.).

SUMMARY

In a sequence of seismic design procedure for tall buildings over 45 metres in Japan, a feed-back dynamic design system has been accepted as the most advanced and reliable one. Dynamic analysis with a rigorous computer program is indispensable to the design procedure as are earthquake observations, structural tests and post construction study. Developments of these major links in the system have made the system itself progress, and we could therefore predict the degrees of damages with full accuracy.

The dynamic analysis, however, requires too much expense and high technique to be utilized for common buildings. It is, then,hoped to establish a new earthquake regulation which provides sufficient safety for buildings.

Both the Tokachi-Oki and the San Fernando earthquakes revealed the defects of earthquake regulations. Much research and investigation proved the lack of deformability and ductility of the structures designed by the codes, causing the destructive damages of buildings. In reply to strong demands for code changes, many people and institutions concerned have been making efforts both in Japan and in the United States.

The author supposes that the new codes would have to define and control the deformability with satisfactory ductility equally as much as the strength of the building structure, and that they would have to be basically applicable to all the earthquake zones in the world. It would indeed be greatly appreciated if all earthquake engineers and scientists could cooperate beyond national bounds for the benefit of mankind.

STOCHASTIC RESPONSE OF STRUCTURES TO EARTHQUAKE EXCITATIONS

by Joseph Penzien

Professor of Structural Engineering and
Director, Earthquake Engineering Research
Center, University of California, Berkeley

STOCHASTIC MODELLING OF STRONG GROUND MOTIONS A)

Since seismic waves are initiated by irregular slippage along faults followed by numerous random reflections, refractions, and attenuations within the complex ground formations through which they pass, stochastic modelling of strong ground motions seems very appropriate. If unlimited ground motion data were available, representative stochastic models could be established directly by statistical analysis. Unfortunately, strong motion data in the form of accelerograms are limited. Therefore, one is forced to hypothesize forms of models and to use the available strong motion data primarily in checking the appropriateness of these forms.

Stationary "White Noise"

Considering the irregular manner in which slippage undoubtedly occurs along the fault, strong ground motions at some distance from the fault might be considered as the superposition of short duration random pulses arriving randomly in time. Therefore, since accelerograms usually have a phase of nearly constant intensity during the period of most severe oscillation, one might consider modelling this phase with a "white noise" process of limited duration. Housner,[1] Rosenblueth,[2] Bycroft,[3] Thompson,[4] and others considered this possibility in their earlier investigations.

For analytical purposes one may wish to generate sample functions which approach "white noise". This procedure can be carried out digitally by first sampling a sequence of pairs of statistically independent random numbers x_1, x_2, x_3, x_4; ... ; x_{n-1}, x_n, all of which have a uniform probability distribution over the range $0 < x < 1$. A new sequence of pairs of statically independent random numbers y_1, y_2, y_3, y_4; ... ; y_{n-1}, y_n are then generated using the relations

$$y_i = (-2 \log_e x_i)^{\frac{1}{2}} \cos 2\pi x_{i+1}$$

$$(1)$$

$$y_{i+1} = (-2 \log_e x_i)^{\frac{1}{2}} \sin 2\pi x_{i+1}$$

which can easily be shown to posess a Gaussian distribution with a mean of zero and a variance of unity.

A sample function $a_r(t)$ can now be established by assigning the values y_1, y_2, ... , y_n to n successive ordinates spaced at equal intervals $\Delta\varepsilon$ along a time abscissa and by assuming a linear variation of ordinates over each interval. Usually, the initial ordinate y_0 is assumed equal to zero and is located at $t = t_0$ where t_0 is a random variable having a uniform probability density function of intensity $1/\Delta\varepsilon$ over the interval $0 < t_0 < \Delta\varepsilon$.

A complete ensemble of m such sample functions $a_r(t)$ (r = 1, 2, ..., m) can be obtained by repeating this procedure, thereby creating a stationary process characterized by the autocorrelation function

$$(2)$$

$$R_a(\tau) = \begin{cases} \dfrac{2}{3} - \left(\dfrac{\tau}{\Delta\varepsilon}\right)^2 + \dfrac{1}{2}\left(\dfrac{|\tau|}{\Delta\varepsilon}\right)^3 & -\Delta\varepsilon < \tau < \Delta\varepsilon \\[2mm] \dfrac{4}{3} - 2\left(\dfrac{|\tau|}{\Delta\varepsilon}\right) + \left(\dfrac{\tau}{\Delta\varepsilon}\right)^2 - \dfrac{1}{6}\left(\dfrac{|\tau|}{\Delta\varepsilon}\right)^3 & \begin{cases} -2\Delta\varepsilon \le \tau < -\Delta\varepsilon \\ \Delta\varepsilon \le \tau \le 2\Delta\varepsilon \end{cases} \\[2mm] 0 & \tau \le -2\Delta\varepsilon \; ; \quad \tau \ge 2\Delta\varepsilon \end{cases}$$

If the intensity of this process is now changed by multiplying each ordinate y_i by the normalization factor $(2\pi S_0/\Delta\varepsilon)^{\frac{1}{2}}$ where S_0 is a constant, the autocorrelation function for the new process becomes

(3)

$$
R_a(\tau) \begin{cases} \left[\dfrac{2\pi S_o}{\Delta\epsilon} \dfrac{2}{3} - (\dfrac{\tau}{\Delta\epsilon})^2 + \frac{1}{2}(\dfrac{|\tau|}{\Delta\epsilon})^3\right] & -\Delta\epsilon \leq \tau \leq \Delta\epsilon \\[3mm] \dfrac{2\pi S_o}{\Delta\epsilon}\left[\dfrac{4}{3} - 2(\dfrac{|\tau|}{\Delta\epsilon}) + (\dfrac{\tau}{\Delta\epsilon})^2 - \dfrac{1}{6}(\dfrac{|\tau|}{\Delta\epsilon})^3\right] & \begin{cases} -2\Delta\epsilon < \tau < -\Delta\epsilon \\ \Delta\epsilon \leq \tau \leq 2\Delta\epsilon \end{cases} \\[3mm] 0 & \tau \leq -2\Delta\epsilon \; ; \; \tau \geq 2\Delta\epsilon \end{cases}
$$

Taking the Fourier transform of Equation (3), one obtains the power spectral density function

$$
S_a(\bar{\omega}) = S_o \left[\frac{12-16 \cos \bar{\omega} \, \Delta\epsilon + 4 \cos 2 \bar{\omega} \, \Delta\epsilon}{(\bar{\omega} \, \Delta\epsilon)^4}\right] \tag{4}
$$

As reported by Ruiz[5] this function is flat to within 5% error for $\bar{\omega}\Delta\epsilon < 0.57$ and to within 10% error for $\bar{\omega}\Delta\epsilon < 0.76$. The function drops to 50% its maximum value S_o at $p\Delta\epsilon = 2$.

It is significant to note that as $\Delta\epsilon$ approaches zero, Equation (3) approaches

$$
R_a(\tau) = 2\pi S_o \, \delta(\tau) \tag{5}
$$

Therefore in the limit as $\Delta\epsilon \to 0$, this process becomes Gaussian "white noise" of intensity S_o over the infinite frequency range $-\infty < \bar{\omega} < \infty$.

Stationary Filtered "White Noise"

Fourier analyses of existing strong motion accelerograms reveal that the Fourier amplitude spectra are not constant with frequency even over a limited band. They are somewhat oscillatory in character, may peak at one or several frequencies, and damp out with increasing frequency all of which suggest that a stationary filtered "white noise" of limited duration could be more representative of actual strong ground motions provided the filter transfer characteristics are properly selected. Kanai[6] and Tajimi[7] have suggested the filter transfer function

$$|H_1(i\overline{\omega})|^2 = \frac{\left[1+4\xi_g^2\left(\frac{\overline{\omega}}{\omega_g}\right)^2\right]}{\left[1+\left(\frac{\overline{\omega}}{\omega_g}\right)^2\right]^2+4\xi_g^2\left(\frac{\overline{\omega}}{\omega_g}\right)^2} \qquad (6)$$

Using this transfer function, the power spectral density function for the filtered process $a_1(t)$ would be

$$S_{a_1}(\overline{\omega}) = |H_1(i\omega)|^2 \, S_a(\overline{\omega}) \qquad -\infty < \overline{\omega} < \infty \qquad (7)$$

where $S_a(\overline{\omega})$ is the power spectral density function for process $a(t)$. Parameters ω_g and ξ_g appearing in the transfer function may be thought of as some characteristic ground frequency and characteristic damping ratio, respectively. Kanai[6] has suggested 15.6 rad/sec for ω_g and 0.6 for ξ_g as being representative of firm soil conditions. Other numerical values should be selected, as appropriate, when significantly different soil conditions are present.

It should be recognized that the above filter attenuates the higher frequency components and amplifies those frequency components in the neighbourhood of $\overline{\omega} = \omega_g$. Since it does not change the amplitudes as $\overline{\omega} \to 0$, some difficulty may arise with the very low frequency components. The cause of this difficulty can easily be recognized by noting that the power spectral density functions for ground velocity and ground displacement are obtained by dividing Equation (7) by $\overline{\omega}^2$ and $\overline{\omega}^4$, respectively. Thus, strong singularities are present at $\overline{\omega} = 0$ which cause the stationary variances of ground velocity and ground displacement to be unbounded. These undesirable singularities can be removed by passing process $a_1(t)$ through another filter which greatly attenuates the very low frequency components. An appropriate filter for this purpose is one having the transfer function

$$|H_2(i\overline{\omega})|^2 = \frac{\left(\frac{\overline{\omega}}{\omega_1}\right)^4}{\left[1-\left(\frac{\overline{\omega}}{\omega_1}\right)^2\right]^2+4\xi_1^2\left(\frac{\omega}{\omega_1}\right)^2} \qquad (8)$$

where the frequency parameter ω_1 and the damping parameter ξ_1 are selected to give the desired filter characteristics. The output process $a_2(t)$ from this filter has a power spectral density function of the form

$$S_{a_2}(\bar{\omega}) = |H(i\omega)|^2 \, S_a(\bar{\omega}) \qquad -\infty < \omega < \infty \qquad (9)$$

where

$$|H(i\bar{\omega})|^2 \equiv |H_1(i\omega)|^2 |H_2(i\omega)|^2 \qquad (10)$$

Equation (9) has the form shown in Figure 1 below.

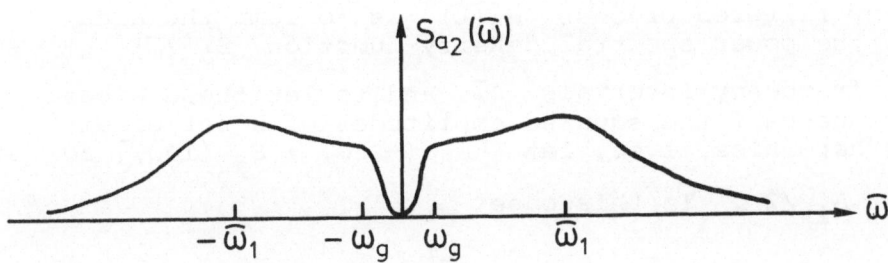

Figure 1: Power spectral density function for filtered stationary "white noise"

The first filtering of process $a(t)$ described above can be accomplished by solving the differential equations

$$\ddot{y}_r + 2\omega_g \, \xi_g \, \dot{y}_r + \omega_g^2 \, y_r = -a_r(t) \qquad (r=1,\, 2,\, ..\, ,m) \quad (11)$$

for y_r and $\dot{y}_r$ using a digital computer and standard numerical integration techniques and then by obtaining a_{1r} using the relation

$$a_{1r} = -2\omega_g \, \xi_g \, \dot{y}_r - \omega_g^2 \, y \tag{12}$$

Likewise, the second filtering can be accomplished by solving the differential equations

$$\ddot{z}_r + 2\bar{\omega}_1 \, \xi_1 \, \dot{z}_r + \bar{\omega}_1^2 \, z_r = -a_{1r} \quad (r=1, \ 2, \ ..,m) \tag{13}$$

for z_r and then by letting

$$a_{2r} = z_r \tag{14}$$

All members of the desired stationary filtered "white noise" process can be obtained by repeating this procedure m times.

A more direct method of obtaining the desired stationary filtered process $a_2(t)$ is to lump the area under the power spectral density function $S_{a_2}(\bar{\omega})$ at equal frequency intervals $\Delta\bar{\omega}$ and to let these areas equal one-half the squared amplitudes of a set of discrete harmonics, i.e., let $[S_{a_2}(-i\Delta\bar{\omega}) + S_{a_2}(i\Delta\bar{\omega})] \, \Delta\omega$ equal $A_{ir}^2/2$. In this case

$$a_{2r}(t) = \sum_i 4 \left[S_{a_2}(i \, \Delta\bar{\omega}) \Delta\bar{\omega} \right]^{\frac{1}{2}} \sin(i\Delta\bar{\omega} \, t + \phi_{ir}) \tag{15}$$

$$(r=1, \ 2, \ .. \ ,m)$$
$$(i=1, \ 2, \ .. \quad)$$

where ϕ is a random phase angle having a uniform probability density function over the range $0< \phi < 2\pi$ .Using this method, $S_{a_2}(\bar{\omega})$ can be expressed in the form of Equation (7) in which case the summation in Equation (15) should be started with $i = h$ where $h\Delta\bar{\omega} = \bar{\omega}_1$. It can also be expressed in the form of Equation (9) in which case the summation in Equation (15) could be started with $i = 1$. This direct method can obviously permit any arbitrary form for $S_{a_2}(\bar{\omega})$ without posing difficulty.

Nonstationary Filtered White Noise

To obtain an even more representative process for strong ground motions, the nonstationary character of actual accelerograms can be considered. Real accelerograms often show a short phase of intensity build-up to some maximum level. This intensity then remains fairly constant for some time after which it decays in an exponential like fashion. This appearance suggests using a nonstationary process $a(t)$ given by

$$a(t) = \Phi(t) \, a_2(t) \tag{16}$$

where $a_2(t)$ is the stationary filtered process previously described and $\Phi(t)$ is an intensity function having an appropriate form. One form which has been suggested is that given in Figure 2. Constants t_1, t_2, and c should be assigned only after considering such factors as earthquake magnitude, epicentral distance, etc.

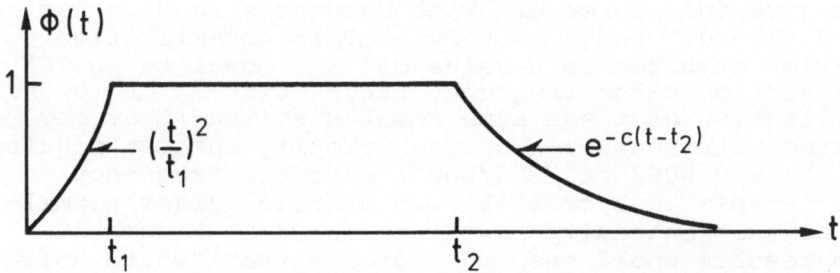

Figure 2: Intensity function $\Phi(t)$ for nonstationary process $a(t)$

EXTREME-VALUE RESPONSE OF 1 DOF SYSTEMS   B)

Stationary "White Noise" Excitation

   Consider the 1 degree of freedom (DOF) linear
system which subjected to earthquake ground motion
$\ddot{v}_g(t)$. Figure 3 shows Housner's pseudo-velocity design spectrum curves for this system for different damping ratios, i.e., $\xi = 0$, 0.02, 0.05, 0.10 . Since
these curves were obtained by normalizing 8 components of recorded ground accelerations (2 components
each of El Centro 1940, El Centro 1934, Olympia 1949,
and Taft 1952) to a common intensity level and by averaging the 8 pseudo-velocity response spectra derived
therefrom, one can consider the ordinates in Figure 3
as representing mean extreme-values of relative pseudo
-velocity. The multiplication factors given in this
figure increase the ordinates to intensity levels corresponding to the earthquake indicated.

   Using an analog computer, Bycroft[3]studied the possibility of using a "white noise" process to represent
earthquake ground motions at a given intensity level.
In these studies, Bycroft noted the extreme-values of
response for a 1 DOF-system using 20 separate
bursts of stationary "white noise" input of 25 seconds
duration each. It was necessary in these studies to
limit the input band with having constant power spectral
density to the range 0-35 cycles per second. To compare his mean extreme-values with Housner's earlier
published velocity spectra, Bycroft normalized his results to that power spectral density of input S_0 which
would give full agreement with Housner's results for
$T_n = 3$ seconds and $\xi = 0.20$. This normalization
criterion resulted in a value of S_0 equal to $0.75 \text{ft}^2/$
cycle/sec over the frequency range $0 < \bar{f} < \infty$. A further
normalization of these same results so that they may be
compared with Housner's design velocity spectra requires
that $S = 0.0063 \text{ ft}^2/\text{rad/sec}^3$ over the frequency
range $-\infty < p < \infty$. Bycroft's mean extreme-values normalized to this intensity level are shown in Figure 3.
These results would seem to indicate that "white noise"
is a reasonable simulation of earthquake ground accelerations.

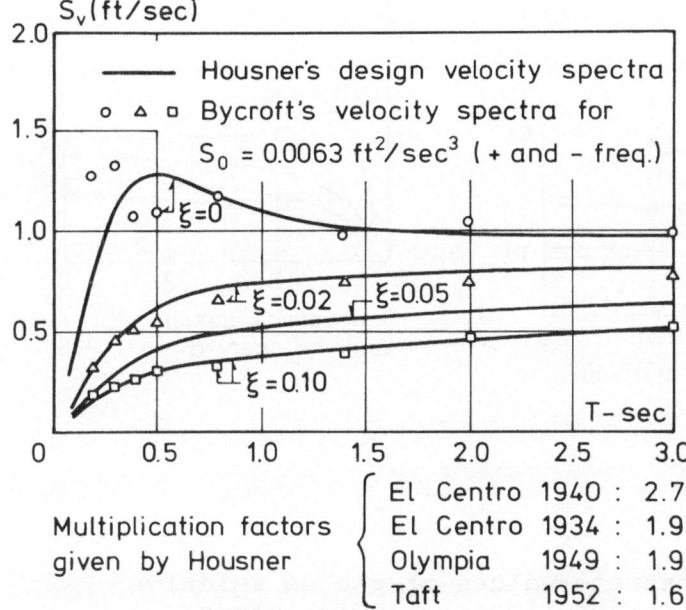

Figure 3: Mean extreme-values of pseudo-relative velocity for linear 1 DOF-systems - Stationary
"White Noise" exication.

Stationary Filtered "White Noise" Excitation

Many investigators have used stationary filtered
"white noise" to simulate earthquake ground accelerations. In one of the investigations, Liu and Penzien [8]
used a single filter having the transfer function given
by Equation (6) with ω_g = 15.6 r/sec and ξ = 0.6 .
Fifty sample functions of band limited "white noise"
were generated by the digital computer methods of Sect.
A with S_o = 0.00614 ft^2/sec^3 and $\Delta\varepsilon$ = 0.025 seconds.
These sample functions, each having 30 second's dura-
tion, were then filtered by digital computer techniques
to provide an ensemble of 50 artificial accelerograms.

Complete time histories of response for the linear
1 DOF-system when subjected separately to each of the
50 input accelerations were established by deterministic methods. The extreme-values of relative displacement were noted in each case and were averaged to obtain mean values. These mean values of displacement

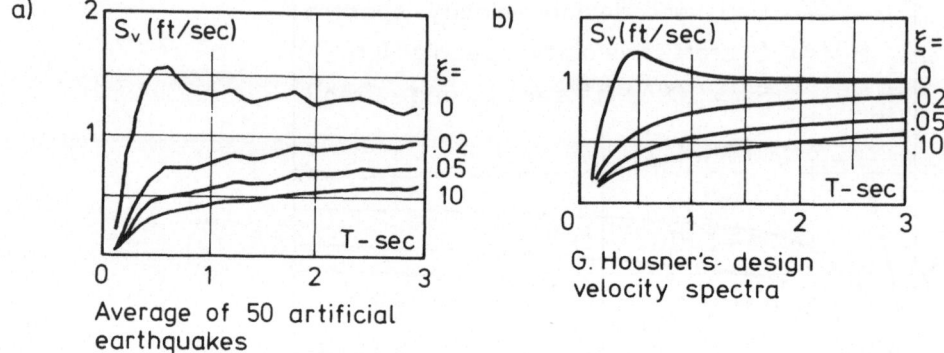

a) Average of 50 artificial earthquakes

b) G. Housner's design velocity spectra

Figure 4:   Mean extreme-values of pseudo relative
            velocity for linear 1 DOF-systems –
            Filtered stationary "White Noise" exci-
            tation.

were then converted to mean extreme-values of pseudo
-velocity by multiplying by ω . These values are
plotted in Figure 4a where they may be compared with
Housner's design spectrum curves in Figure 4b. The
close agreement of these two sets of curves lends support to using filtered stationary "white noise" in the
simulation of strong earthquake ground motions.

Complete time histories of relative displacement
response $v(t)$ for the ordinary elasto-plastic and
stiffness degrading models were established by standard
numerical integration procedures when subjected separately to support accelerations $\ddot{v}_g(t)$ corresponding
to the above described filtered process but after normalizing by a factor of $(2.90)^2$ so that the process
intensity S_O would represent the intensity of the
N-S component of the 1940 El Centro earthquake ($S_O =
(2.90)^2 (0.00614) = 0.01515$ ft^2/sec^3) .

The basic parameters of these nonlinear models,
which are comparable to those used for the linear models,
are shown in Figure 5a.

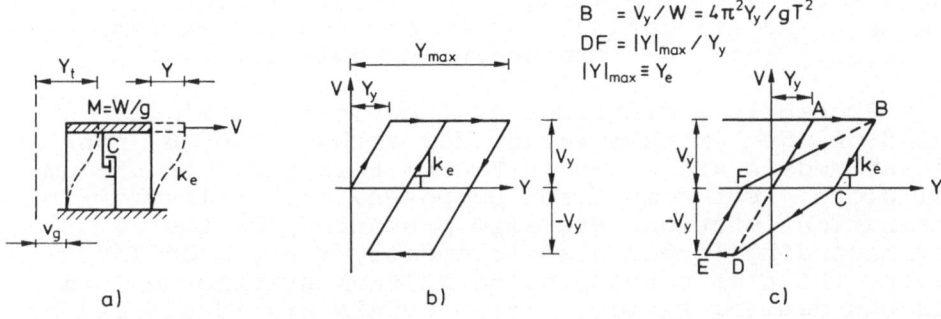

Figure 5: Nonlinear models of 1 DOT-system

In all cases T and ξ represent the period of
vibration and viscous damping ratio, respectively, in
the initial elastic range. The static force-deflection
relations for the elasto-plastic and stiffness degrading models are shown in Figure 5b and 5c, respectively
The strength ratio B and ductility factor DF are defined for these models in accordance with the relations
$B \equiv y_y/W$ and $DF \equiv |y(t)|_{max}/y_y$. It is significant
to note that in addition to loss of stiffness following
any yielding, the stiffness degrading model permits hysteresis loops to be formed even at very low amplitudes
of oscillation. Therefore, this model dissipates more
energy in the lower amplitude ranges of response than
does the equivalent elasto-plastic model.

The response of the elasto-plastic and stiffness
degrading models are considered for two different
periods, $T = 0.3$ and 2.7 seconds, and for two different damping ratios $\xi = 0.02$ and 0.10; thus, the response of 8 different nonlinear models as presented
in Table 1 are discussed. Strength ratios B are
based on the assumption that the yield resistance y_y

equals twice the design load as specified in the 1973 edition of the Uniform Building Code for moment resisting frames, i.e., $B = 2KC = (2)(0.67)(0.05)(T)^{-1/3}$. The mean and standard deviation of the 50 extreme -values in each case are shown in Table 1 .

Probability distribution functions $P(|Y|_{max})$ based on 50 extreme-values for each of the 8 nonlinear models are shown in Figure 6 in the form of Gumbel plots. For comparison purposes, probability distribution functions are also presented for the 4 corresponding linear elastic models, i.e., models having the same corresponding initial stiffnesses and viscous damping ratios. These models are identified by the Arabic numerals 1-12 in Figure 6 and have the properties listed in Table 1 .

Two probability distribution functions are shown in Figure 6 for each of the 12 structural models, namely a wavy line function which is a plot of the actual extreme-values determined for process Y(t) and a straight line function which is the theoretical distribution (Type I) of the form

$$P(|Y|_{max}) = \exp\left[-\exp(-\hat{Y})\right] \qquad (17)$$

where Y is the reduced extreme-value defined by

$$\hat{Y} \equiv \alpha(|Y|_{max}-u) \qquad (18)$$

Case no.	Structural type *	Period T-sec	Damping Ratio-E	Strength Ratio-B	Yield displ. x - IN.	σ_x IN.	$\bar{x}$ IN.	u IN.	$1/\alpha$
1	E	0.3	0.02	–	–	0.115	0.768	0.722	0.085
2	EP	0.3	0.02	0.10	0.088	1.613	3.214	2.450	1.390
3	SD	0.3	0.02	0.10	0.088	0.711	2.480	2.144	0.613
4	E	0.3	0.10	–	–	0.050	0.354	0.330	0.043
5	EP	0.3	0.10	0.10	0.088	0.910	1.947	1.517	0.784
6	SD	0.3	0.10	0.10	0.088	0.360	1.327	1.157	0.310
7	E	2.7	0.02	–	–	3.07	14.15	12.73	2.59
8	EP	2.7	0.02	0.048	3.42	5.51	16.35	13.75	4.75
9	SD	2.7	0.02	0.048	3.42	5.83	14.32	11.56	5.02
10	E	2.7	0.10	–	–	1.31	8.77	8.24	0.97
11	EP	2.7	0.10	0.048	3.42	4.56	11.57	9.41	3.94
12	SD	2.7	0.10	0.048	3.42	3.26	9.98	8.45	2.80

* E - Elastic
EP - Elasto-Plastic
SD - Stiffness Degrading

Table 1: Statistical properties of the response parameters

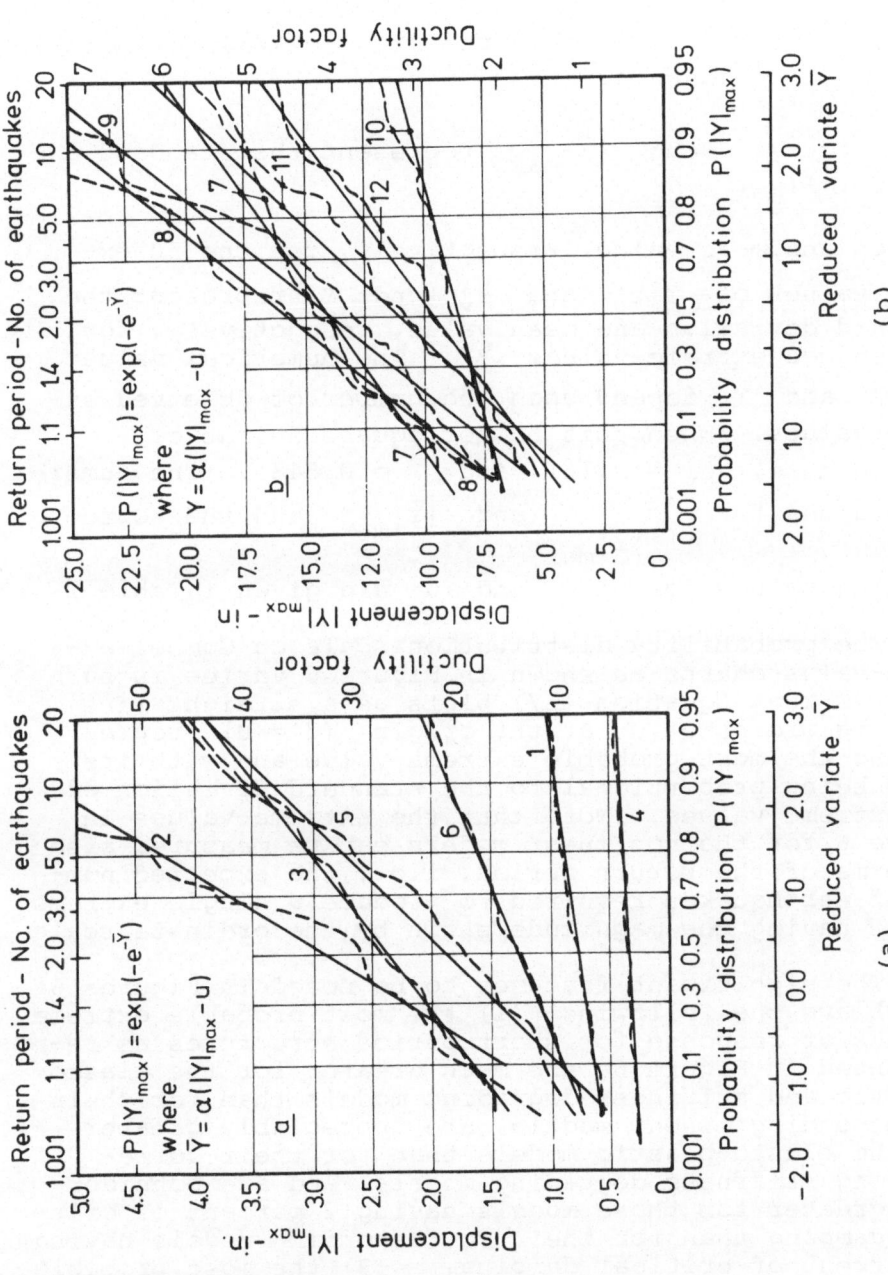

Figure 6: Probability distribution for extreme values of relative displacements

Constants α and u can be determined using the relations

$$1/\alpha = \sigma_{|Y|_{max}} / \sigma_{\hat{Y}} \quad ; \quad u = |\bar{Y}|_{max} - (\hat{\bar{Y}}/\alpha) \tag{19}$$

where $\sigma_{|Y|_{max}}$ and $|\bar{Y}|_{max}$ represent the standard deviation and mean value, respectively, for the 50 extreme-values of $Y(t)$ and $\sigma_{\hat{Y}}$ and $\hat{\bar{Y}}$ represent the standard deviation and mean value, respectively, for the reduced extreme-values $\hat{Y}$. The numerical values of $\sigma_{\hat{Y}}$ and $\hat{\bar{Y}}$ depend upon the number of observed extreme-values. When this number equals 50, as considered here, $\sigma_{\hat{Y}} = 1.161$ and $\hat{\bar{Y}} = 0.548$. The numerical values for $\sigma_{|Y|_{max}}$ and $|\bar{Y}|_{max}$ and the corresponding values for $1/\alpha$ and u are given i Table 1 .

The probability distribution scale on Gumbel extreme-value charts as shown in Figure 6 varies in such a manner that Equation (17) plots as a straight line with its ordinate u at the origin $(\hat{Y} = 0)$ representing the most probable extreme-value and with its slope being proportional to the standard deviation of the extreme-values. Note that the extreme-values in Figure 6 for the nonlinear models can be measured also in terms of the return period, i.e., the expected number of earthquakes required to produce a single extreme-value having the magnitude shown by the ordinate scale.

The significant features to be noted in Figures 6a and 6b are the following: (1) the most probable extreme-values of response for short period structures as represented in Figure 6a are much greater for the elasto-plastic and stiffness degrading models than for their corresponding linear models, are appreciably greater for the elasto-plastic models than for their corresponding stiffness degrading models, and are considerably greater for those models having 2 percent of critical damping than for their corresponding models having 10 percent of critical damping. (2) the most probable extreme-values of respone for long period structures as represented in Figure 6b are considerably greater for those models having 2 percent of critical damping than for their corresponding models having 10 percent of critical damping; however, these values differ very little from one model to another. (3) the standard deviations of extreme-value response for the short period structures are considerably larger for the elasto-plastic and stiffness degrading models than for their corresponding linear models and are appreciably larger for the elasto-plastic models than for their corresponding stiffness degrading models. (4) the standard deviations of extreme-value response for long period structures correlate in a manner quite similar to short period structures except that the differences are not so great. (5) increasing the viscous damping ratio increases the standard deviations of extreme-value response for each model type. (6) the theoretical straight line functions as represented by Equation (17) shows very good correlations with the actual distribution

The probability distribution functions for extreme-values shown in Figure 6 result from an input process $\ddot{v}_g(t)$ having a duration of 30 seconds. The corresponding extreme-values will, of course, be less for processes of shorter duration. To illustrate these effects, a ratio of the ensemble average of extreme-values for an input process of duration T_O to the ensemble average of extreme-values for an input process of 30 seconds is plotted in Figure 7 as a function of the duration ratio $T_O/30$.

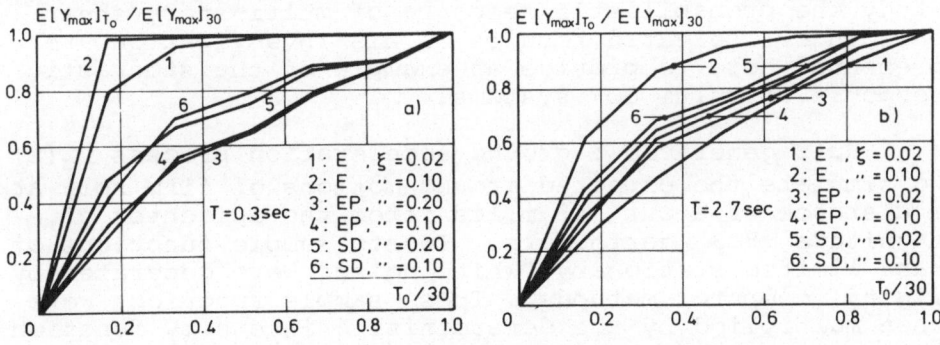

Figure 7: Duration effect of stationary process on mean peak response of linear and nonlinear structures.

262

It is quite evident, based on Curve No.2 in Figure
7a, that the mean peak response of typically damped,
linear, short period structures (T = 0.3 seconds) increases very slowly with duration beyond approximately
6 seconds. Long period structures are, of course, more
sensitive to duration as shown by Curve No.2 in Figure
7b. This curve indicates that the magnitude of mean
peak response for a 15-second duration process is approximately 95 percent of the magnitude observed for
a 30-second duration period. As shown in Figures 7a
and 7b, elasto-plastic and stiffness degrading structures are much more sensitive to duration than are
elastic structures; thus, it is apparent that realistic durations must be used for stationary inputs when
investigating the response of nonlinear structures.

As demonstrated above, stationary processes of
short duration can be used quite effectively to establish the probabilistic peak response of both linear and
nonlinear systems to strong motion earthquakes of a
given intensity level. However, as the true dynamic
characteristics of real structures become better known,
damage will likely be measured using various accumula-
tive damage criteria in which case it may be desirable
to use appropriate nonstationary processes for the excitation.

EXTREME VALUE RESPONSE OF MULTI DOF SYSTEMS   C)

Numerous investigators have suggested the use of
nonstationary processes to represent strong ground motions. One such process was established by Ruiz[5] to
study the probabilistic response of multi-story shear
buildings. Selected results of his investigation are
presented here to provide an example of the stochastic
response of multi DOF systems.

Ruiz generated a ground acceleration process $\ddot{v}_g(t)$
to simulate the expected ground motions of firm soil at
a distance of about 45 miles from the epicenter of a
magnitude 8.3 earthquake. Twenty sample functions of
band limited stationary "white noise" were generated by
digital computer methods. These sample functions were
then multiplied by the deterministic intensity function
$\Phi(t)$ shown in Figure 2 with $t_1 = 0$, $t_2 = 11.5$ sec.,
and $c = 0.155$ sec^{-1} . The resulting nonstationary wave
forms were then filtered once using the methods described in Sect.A with the filter transfer function
being of the form given by Equation (6) with $\omega_g = 15.7$

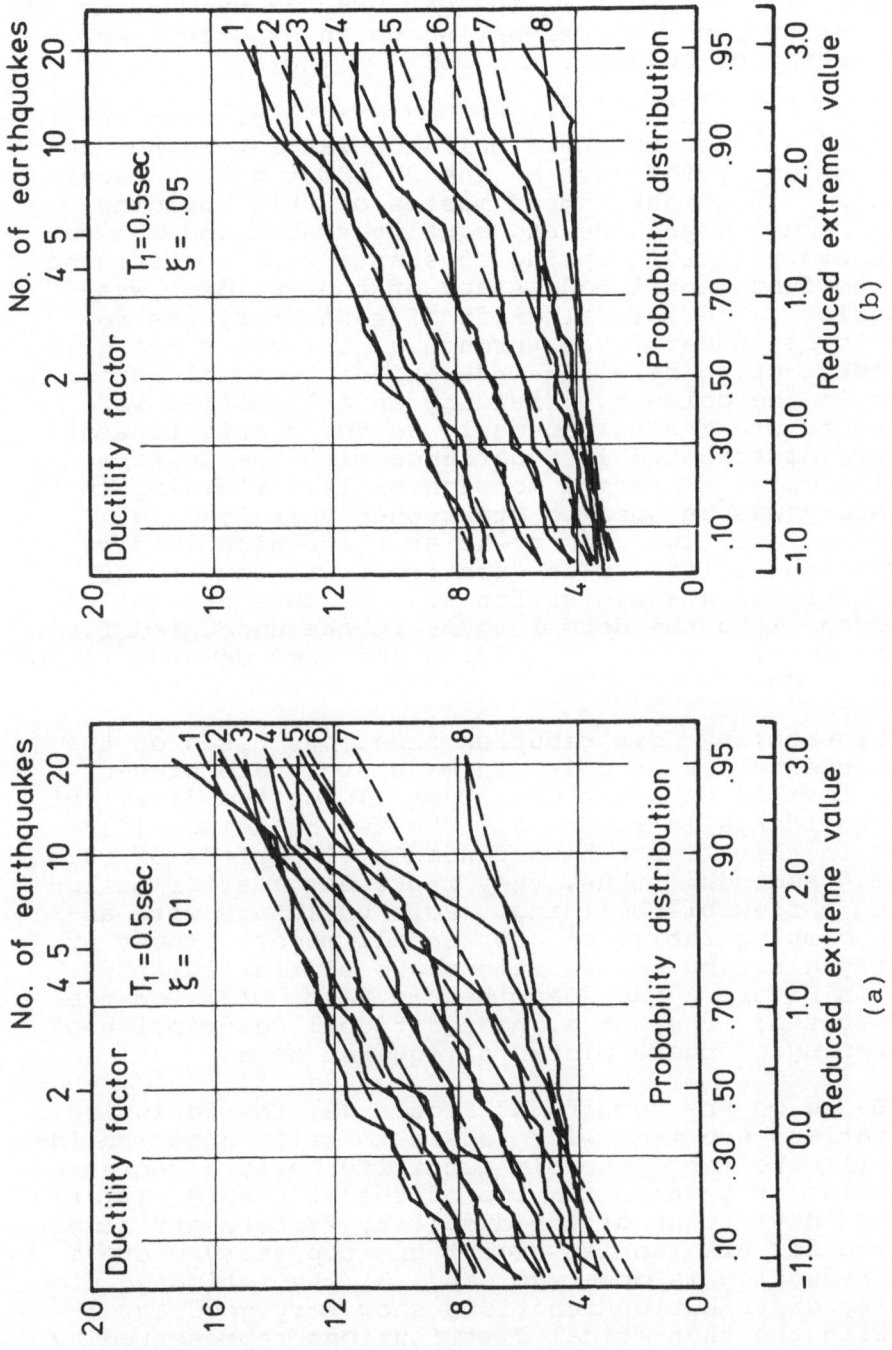

Figure 8: Probability distribution for storey ductility factors

rad/sec and $\xi_g = 0.6$. The process was normalized to an intensity level corresponding to an expected peak acceleration of 0.3 g .

The complete time history of elasto-plastic re-sponse of an 8 story shear building was determined de-terministically for each of the 20 input ground accel-erations. The eight lumped masses of this building were of equal magnitude and equally spaced and the re-lative story elastic spring constants were adjusted so that the fundamental mode shape of the building was triangular. The lateral drift of each story was re-lated to its shear force through a bilinear hysteretic force-deflection relation independent of axial forces acting in the columns. Yielding in all stories were assumed to start simultaneously as the static lateral loading, distributed in accordance with the Uniform Building Code[9], increased monotonically. Yielding in each story was assumed to start when this loading reached a level twice as great as the design loading. The yielding stiffness in each story was set at 10% of its initial elastic stiffness. Viscous damping was introduced into the normal modes in the uncoupled form and was controlled by specifying the same damping ratio in each mode.

Probability distribution functions based on the 20 extreme-values of drift in each story are present in the form of Gumbel plots (Type I) for two different shear buildings in Figure 8. The two buildings iden-tified in this figure have fundamental periods of 0.5 and 2.0 seconds; thus, they represent a stiff build-ing and a flexible building. Both buildings were as-signed damping ratios of 5% in all modes. These pro-bability distribution functions are similar to those shown in Figure 6 and described in Sect. B(2) for the 1 DOF-system; therefore, no additional description of the meaning of these plots is required here.

Based on the results of Figure 8a, the following observations are made with regard to stiff shear build-ings: (1) the most probable ductility factors decrease monotonically towards the top of the structure, (2) the standard deviations of the ductility factors are almost the same for all stories except the top story where a large reduction is observed, and (3) the estimated pro-bability distribution functions show very good agree-ment with the theoretical distributions represented by the straight lines. Likewise based on the results in Figure 8b, the following observations are made with re-

gard to flexible shear buildings: (1) the most probable
ductility factor decreases towards the middle stories
and then increases towards the top stories reaching a
value in the top story of comparable magnitude with
that in the first story, (2) the standard deviations of
the ductility factors decrease towards the upper
stories but with a slight increase in the top story,
and (3) the agreement between the estimated probability
distribution and the theoretical straight line distribution is not as good as for the stiff shear building.
Comparing Figures 8a and 8b, it is clear that the most
probable ductility factors and their standard deviations are higher for the stiffer structures.

REFERENCES

1. Housner, G.W., Characteristics of strong-motion
       earthquakes, Bull. Seism. Soc. Amer., 37, 19,1947.

2. Rosenblueth, E., Some applications of probability
       theory in aseismic design, 1st World Conf. Earthq.
       Eng., Berkeley, Cal., 1, 8-1, 1956.

3. Bycroft, G.N., White noise representation of earth-
       quakes, Proc. ASCE, Journal of the Eng. Mech.Div.,
       86, 1, 1960.

4. Thompson, W.T., Spectral aspects of earthquakes,
       Bull. Seism. Soc. Amer., 59, 91, 1959.

5. Ruiz, P. and Penzien, J., Probabilistic study of
       the behaviour of structures during earthquakes,
       Earthq. Eng. Res. Center Rep., 69-3, Univ.Cal.
       Berkeley, 1969.

6. Kanai, K., Semi-empirical formula for the seismic
       characteristics of the ground, Bull. Earthq.Res.
       Inst. Univ. Tokyo,35, 309, 1957.

7. Tajimi, H., A statistical method of determining the
    maximum response of a building structure during
    an earthquake, 2nd World Conf. Earthq. Eng.,Tokyo
    & Kyoto, 2, 781, 1960.

8. Penzien, J. & Liu, S.C., Nondeterministic analysis
    of nonlinear structures subjected to earthquake
    excitations. 4th World Conf. Earthq. Eng.Chile,
    1, A-1, 114, 1969.

9. Uniform Building Code, International Conference of
    Building Officials, Section 2313(d), 1967.

CODES AND REGULATIONS: PROBLEM OF IMPLEMENTATION

by Aybars Gürpinar
Assistant Professor, Dept. of Engin. Sciences,
Middle East Technical University, Ankara.

INTRODUCTION

Research related to engineering seismology and earthquake engineering has been increasing at a very rapid rate since the late 1950's. This is partly due to the variety of the disciplines included in the subject and partly because of its immediate applicability to concrete problems. The Fifth World Conference on Earthquake Engineering in Rome is an example of the amount and nature of the work that is being done in all parts of the world. However, it is difficult to understand the quality and the quantity of the work being done on the subject simply by looking at the building codes and regulations of earthquake-prone regions of the world. Although the amount of research being done is immense, the code makers are very cautious and hesitant about incorporating such research into building codes and they are usually satisfied with an approach that is more pragmatic.

As a simple example, stochastic methods have been applied to this subject very frequently, yet no building code reflects this type of research. In Earthquake Engineering Research published by the National Research Council (1969) the relationship between pure research, applied research and practical application with regard to earthquakes has been summarized (Figure 1). Practical application may be regarded as a subset of applied research which itself is a subset of pure research.

268

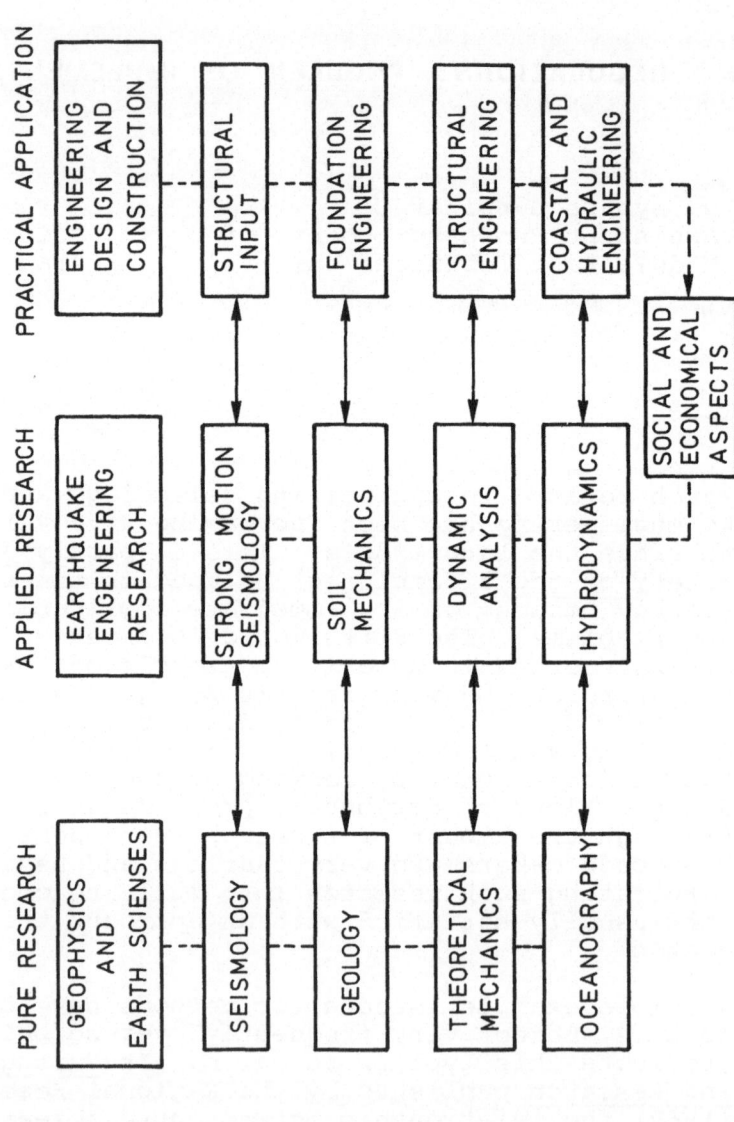

Figure 1: Relations Between Pure Research, Applied Research, and Practical Application

There are two basic problems in the relationship between applied research and practical application. The first one is the reflection of the new developments in earthquake engineering research in building codes and regulations. This problem is almost universal and results from the dilemma of the function of a code. The second basic problem involves the implementation of the existing codes; in other words, the problem is not one of the contents of the code, but its applicability. This problem is an acute and chronic one especially for developing countries. Although it is this problem that is responsible for a great majority of the casualities due to earthquakes, very little is done about it because it has a nebulous, amorphous and political character and the engineer feels arkward and uneasy in handling it.

DESIGN OF A BUILDING CODE

The problem concerning the contents of building codes can be approached systematically by a 'design tree' technique. [1] A design tree consists of questions and answers as shown in Figure 2. The tree is started with a statement of the primary problem, which is referred to as the zeroeth order Q_0. The alternative solutions, first order, are represented by slanting

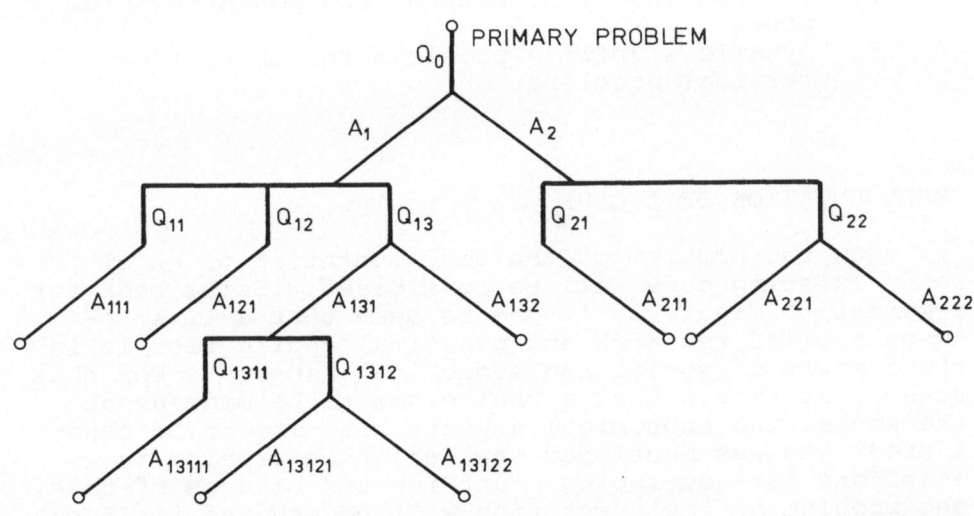

Figure 2: Design Tree

lines A_i . For each of these there may be several
problems to be solved. Any one or more of the alternative solutions is sufficient to solve the problem. On
the other hand, it is necessary to solve all the problems before a lower order alternative solution is obtained. The primary question Q_O in this case is
"what constitutes an aseismic design code?". The first
order alternatives may be A_1 : equivalent static force
approach and A_2 : direct approach. A_1 is the summary
of the different ways of handling the problem by various codes used by different countries (Figure 3). A_2,
on the other hand, is a summary of the more direct approach (Figure 4). This approach separates the problem into three major parts: excitation; system; and
response limitation. Explanation of the symbols of
Figures 3 and 4 can be seen in Table 1. Any new technique can be easily incorporated into the scope of this
approach by adding an alternative solution. By comparing A_1 and A_2 , the shortcomings of the existing
codes can be easily observed. These can be summarized
as follows:

1. Energy dissipation characteristics of structures
2. Inelastic behaviour of structures
3. Low cycle fatigue behaviour
4. Random vibration methods
5. Reliability considerations
6. Soil-structure interaction
7. Interaction of structural and non-structural
   parts
8. Dynamic stability problems resulting from
   vertical acceleration

IMPLEMENTATION OF A CODE

Now the problem of the implementation of an already existing code will be considered. Going back for
a moment to Figure 1, it can be seen that a link between applied research and practical application is in
the context of social and economical aspects. For this
reason, at this point, a choice has to be made about
the social and economical aspects that are to be considered. As was mentioned earlier, this problem is an
acute one for developing countries and because of this,
the problem of implementation will be studied for rural
dwellings in Eastern Turkey. With a few local modifications it can be generalized for the rural parts of
most other developing countries.

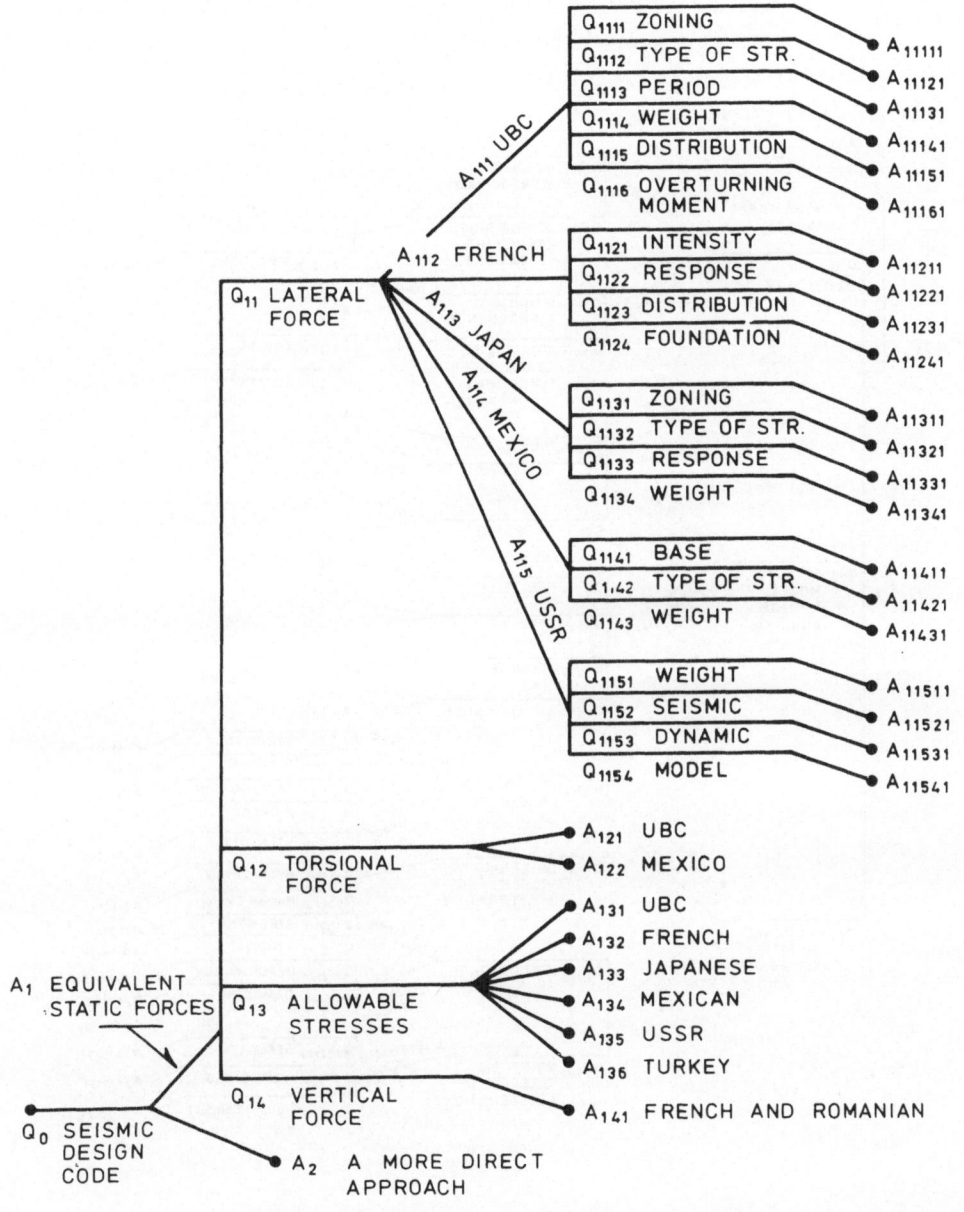

Figure 3: "Design Tree" for Existing Seismic Design Codes

272

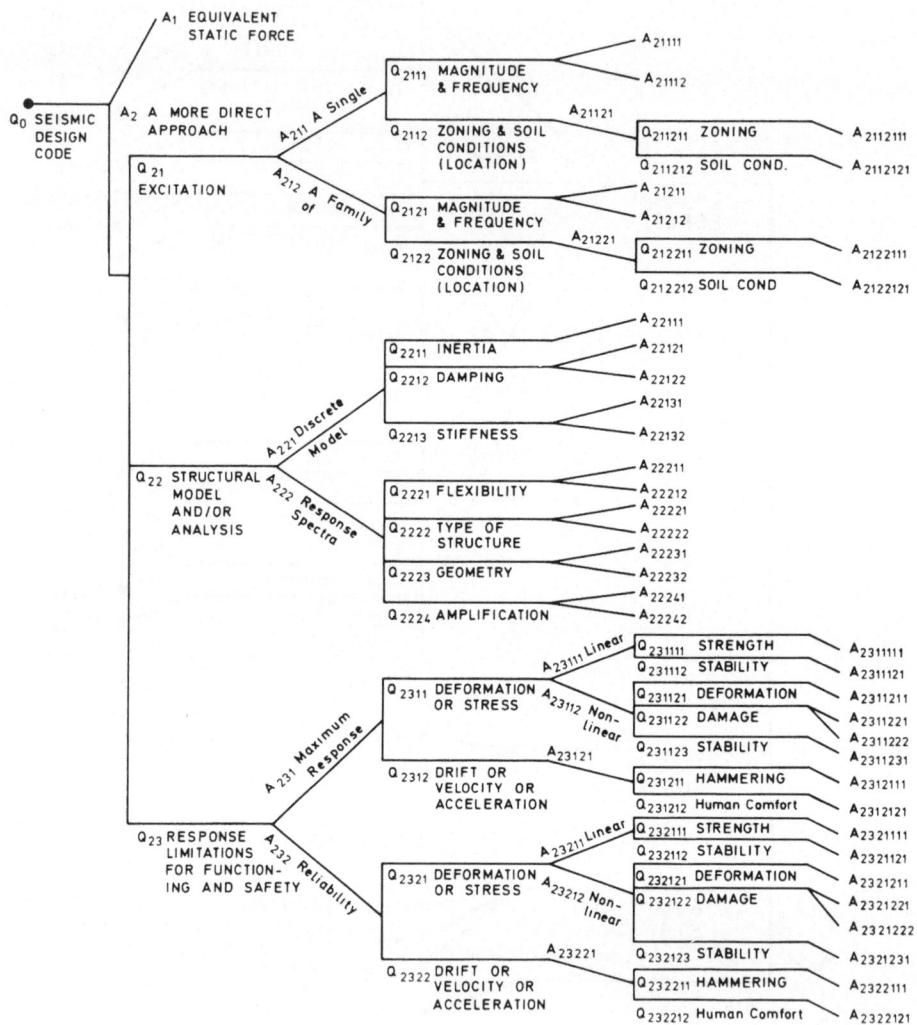

Figure 4: "Design Tree" for a Direct Approach

Table 1

"Answers" in Design Trees as Shown in Figures 3 and 4

A_{11111}:

Zone No.	0	1	2	3
Z	0	.25	.50	1.0

A_{11121}: $K =$ .67   for Ductile Frames
            1.33   for Box Systems (e.g.,shear walls)
            3.0   for Elevated Tanks

A_{11131}: $C = \dfrac{.05}{\sqrt[3]{T}} \le 0.1; \quad T = \begin{cases} 0.1 \ N, \text{ for Ductile Frames} \\ \dfrac{.05}{\sqrt{D}} \ h, \text{ otherwise} \end{cases}$

A_{11141}: $W =$ total weight

A_{11151}: $F_f = 0.004 \ V \dfrac{h_n^2}{D_s} \le .15 \ V$, top of structure

   $F_x = \dfrac{(V - F_t) w_x h_x}{\sum\limits_{i=1}^{n} w_i h_i}$, otherwise

A_{11161}: $M = J(F_t h_N + \sum\limits_{i=1}^{n} F_i h_i), \quad J = \dfrac{.5}{\sqrt[3]{T^2}} \le 1.0$

   $M_x = J_x [F_t(h_N - h_x) + \sum\limits_{i=x}^{N} F_i(h_i - h_x)]$,

   $J = J + (1-J) \left(\dfrac{h_x}{h_N}\right)^3$, for other floors

A_{11211}:

MMI	VIII	IX	X	XI	XII
	1	2	4	8	16

A_{11221}: $.05 \le \beta = \dfrac{.065}{\sqrt[3]{T}} \le .10$; Common Buildings (norm-
                  al damping)

Table 1 (continued)

$$.65 \le \beta = \frac{.085}{\sqrt[3]{T}} \le .13;$$ Large Span and Little Partitions (medium damping)

$$.06 \le \beta = \frac{.095}{\sqrt[4]{T^3}} \le .20;$$ Skeleton, no external friction (low damping)

A_{11231}:
$$\sigma_x = \frac{\sum_{i=1}^{x} M(z_i)q(z_i) + \int_{0}^{h_x} m(z)q^2(z)dz}{\sum_{i=1}^{x} M(z_i)q^2(z_i) + \int_{0}^{h_x} m(z)q^2(z)dz} \, q(h_N) \, \frac{\Gamma(T)}{g}$$

$q(h)$ = shape fcn. of the fundamental mode,
$\Gamma(T)$ = max. response (acceleration of a SDF system with the same period T and damping as the fundamental mode. (The first three modes are considered for slender structure with low damping)

A_{11241}: $f = 0.8$ for deep foundation of firm rock
$= 1.3$ for piling in moist ground

A_{11311}: $z = 0.8$ or 0.9 depending on seismic zone

A_{11321}: $s = 0.6$ for steel frame or wooden structure on firm ground,
$= 0.9$ for semi-firm ground and reinforced concrete structure,
$= 1.0$ for soft ground

A_{11331}: $c = 0.2$ for structures up to 16 m., and add 0.01 for each 4 m. in addition to 16 m.

A_{11341}: W = total weight

A_{11411}: Type 1 = structures having, at right angles, with the direction being considered, two or more elements capable of resisting shear, and whose deformation under lateral loads is essential due to flexure in members.

Table 1   (continued)

Type 2 = Structures whose deformations under
lateral loads are due to shearing
stresses or axial forces in members.

Type 3 = Elevated tanks, shimney stacks,etc.

Type \ b \ Zone	1	2
1	.06	.04
2	.08	.08
3	.15	.10

A_{11421}:   k = 1.3   for Group A, government, municipal
and public buildings

= 1.0   for Group B, private housing

0   for Group C, isolated and unimport-
ant buildings

A_{11431}:   W = total weight

A_{11511}: Z_q= D.L.+ 0.8 L.L. (for warehouses, use full
L.L.)

A_{11521}:

GEOFIAN Scale	7	8	9
α	1/40	1/20	1/10

4 Types of Buildings:

I: Massive buildings of great importance,
increase rating by 1

II: Public buildings, industrial buildings of primary importance, same rating

III: Industrial buildings of secondary importance, one storey houses, same ratings for "6", "7", reduce "8" and
"9" by 1

IV: Use rating "6" for all farm buildings,
barns, sheds, etc.

A_{11531}: $0.6 < \delta = \frac{0.9}{T} < 3.0$, and use 1.6 δ for
slender and flexible struct-
ures

Table 1 (continued)

A_{11541}:

$$\zeta_x = \frac{q(x_k) \sum\limits_{i=1}^{N} Q_i q(x_i)}{\sum\limits_{i=1}^{N} Q_i q^2(x_i)}$$

use 1st mode for normal buildings, and use the first 3 modes for slender and flexible buildings

A_{121}:

$$T_N = S_N(1.5 \, e_N + 0.5 \, B_n)$$

shear eccentricity max. dimension

A_{122}:

$$e_D = 1.5 \, e_c \pm 0.05 \, D$$

A_{131}: May be increased by 33%

A_{132}: May be increased by 50%

A_{133}: May be increased by 50% for reinforcing steel, and 100% for concrete in reinforced concrete

A_{134}: May be increased by 50% for steel, and 33% for concrete

A_{135}: May be increased by 40% for steel, 20% for reinforced concrete, and 0% for pre-stressed concrete

A_{136}: May be increased by 50% for good soil conditions, 30% for fair soil conditions, and 0% for bad soil conditions

A_{141}: French and Romanian codes mentioned it. However, it is not clear as to how to do it in designs

A_{21111}: Take a typical set of earthquake records, e.g., 1940 El Centro.

A_{21112}: Find the least fovourable excitation, e.g., Drenick

Table 1   (continued)

$A_{21112111}$:   Zoning (deterministic)

$A_{21112121}$:   Soil Conditions (deterministic)

A_{21211}:   Take a set of artificial functions, e.g., Shinozuka and Sato, Amin and Ang, etc.

A_{21212}:   Take several sets of artificially generated functions, e.g., Jennings, Housner and Tsai

$A_{2122111}$:   Zoning (statistical)

$A_{2122121}$:   Soil Conditions (statistical)

A_{22111}:   Analytical

A_{22121}:   Analytical

A_{22122}:   Experimental

A_{22131}:   Analytical

A_{22211}, A_{22221}, A_{22231}, A_{22241}:   Linear-Discrete Mechanics *

---

* discrete mechanics, based on the calculus of finite differences, can be used to obtain field or functional solutions for systems, which are more accurately represented by a lattice or a network of elements.

A_{22212}, A_{22222}, A_{22232}, A_{22242}:   Nonlinear analysys

$A_{2311111}$: $|s_{max}| \leq s_y$

$A_{2311121}$:   Stable

$A_{2311211}$: $|y_{max}| \leq y_{limit}$

$A_{2311221}$: $\sum D_i \leq 1$

Table 1   (continued)

$A_{2311222}$:   Energy Absorbed $\leq E_{crit}$

$A_{2311231}$:   Stable

$A_{2312111}$:   Hammering Prevented

$A_{2312121}$:   Comfortable

$A_{2321111}$: $P(|s_{max}| \leq s_y) \leq P_1$

$A_{2321121}$: $P(\text{Unstable}) \leq P_2$

$A_{2321211}$: $P(|y_{max}| \geq y_{limit}) \leq P_3$

$A_{2321221}$: $P(\sum D_i \geq 1) \leq P_4$

$A_{2321222}$: $P(E_{ab} \geq E_{cr}) \leq P_5$

$A_{2321231}$: $P(\text{Unstable}) \leq P_6$

$A_{2322111}$: $P(\text{Hammering}) \leq P_7$

$A_{2322121}$: $P(\text{Discomfort}) \leq P_8$

P_i's   should be functions of type of structure, function (importance) of structure, consequence of failure, material properties, etc.

Table 1:   Explanation of Symbols Used in Figures 3 and 4.

The implementation of the code is a difficult problem in rural areas. Here the dwellings are usually built by the owner or at best by local masons. These masons may be experienced builders but usually they are totally ignorant of earthquake-resistant design. What is more is that some standard dwelling designs for rural areas may prove to be disastrous when subjected to earthquake-induced forces. As an example, the plan of a typical eastern Anatolian dwelling is shown in Figure 5. Here one or two other units attached to the dwelling can be seen. These are the barns, one for the cattle and the other for the sheep. The dwelling part is usually designed and constructed with a fair amount of care, but the barns are built vary carelessly. (Figure 6). The unfortunate fact, however, is that these structures are attached to each other and function jointly when subjected to lateral forces. In other words, the collapse of the barns means the collapse of the dwelling. Another disadvantage of this seemingly harmless attachment of the barn is that it destroys one of the few desirable properties of rural dwellings with respect to earthquake-resistant behaviour, and that is symmetry.

In Eastern Turkey dwellings are designed not particularly to resist earthquakes but to protect the inhabitants from the bitter cold winters of the region.

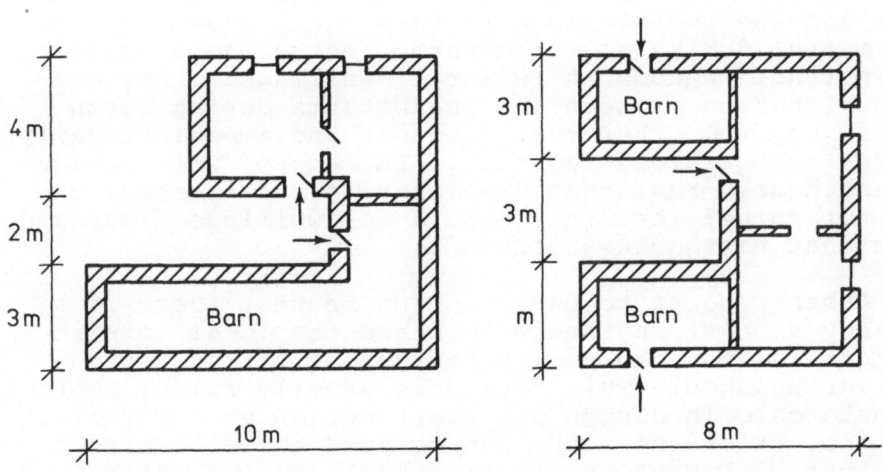

Figure 5: Layout of Typical Eastern Anatolian Dwellings

Figure 6a: Typical Barn Structure in Rural Areas

In fact, the attachment of the barns to the dwelling is for the purpose of using the heat generated by the animals. Another feature that is used for protection from the cold is the excessive thickness (60 cm) of the walls and the roofs. This feature again proves to be detrimental during an earthquake. These facts point to an important phenomenon in rural dwellings: they are not designed to protect the inhabitants during earthquakes but from other more frequent and immediate daily menaces such as cold and heat. Therefore, when designing earthquake-resistant dwellings in rural areas, one must not forget that they should be dwellings first and after that earthquake-resistant.

Another point to bear in mind is that there is a tendency amongst engineers to blame the local construction material for the poor behaviour of structures during earthquakes. This notion is usually false. With adequate care in design and construction such materials as stone, brick and adobe can be used to build structures that will perform satisfactorily during earthquakes. In Figure 7 photographs of structures built by local masons which have endured several earthquakes can be seen.

Figure 6b: Typical Barn Structure in Rural Areas

Last but not least, the importance of construction
quality must be stressed. Figures 8-9 show photographs
of a housing project developed after an earthquake in
1963 in a town called Tuzluca. These dwellings were
designed by government offices and their construction
was carried out by a contractor.

Due to inferior construction most of the dwellings
are still empty. It should be pointed out that no destructive earthquake occurred since 1963 in the vicinity of this town.

In the light of the above examples, the only vi-
able possibility seems to be to limit the scope of
the aseismic code to engineer-built structures. The
vast problem of rural construction should be solved by
public education. A practical suggestion might be to
open up training courses for masons providing them with
various ready-made typical designs. The Ministry of
Reconstruction and Resettlement of Turkey is already
taking some measures along the line of public education.

It should be remembered that the rural dwellings,
at this stage, a small increase in cost results in a
substantial decrease in seismic risk.

Figure 7: Good-Quality Masonry Structures of Proved
Earthquake Resistance

Figure 8: Post-Earthquake Housing Project in Tuzluca, Turkey

Figure 9: Foundation Detail

REFERENCES

1. Yao, J.T.P., Omidvaran, C., Gürpinar, A.,.Hulsbos, C.L., Seismic Design of Building Structures", Technical Report S-10, Construction Engineering Research Laboratory, Corps of Engineers, July 1972.

AN EPIDEMIOLOGIST'S VIEW OF EARTHQUAKES

by Michel F.Lechat, M.D., Dr.P.H.
Universite Catholique de Louvain,
Ecole de Sante Publique,
Bruxelles.

EARTHQUAKES AND EPIDEMIOLOGY

There are many ways of looking at the social and health effects of earthquakes, or of natural disasters in general, each corresponding to a definite objective.

One is to view natural disasters as large-scale unplanned ecolocical experiments, providing an opportunity to investigate the psycho-social reactions of populations. Following earthquakes, large numbers of people are abruptly stripped of the things and structures which support their life. They are thrown back with little or no warning or preparedness to primitive evolutionary conditions, fighting for survival, food, shelter and protection of relatives. Natural disasters make therefore a powerful tool to study basic behavioural patterns or attitudes, such as allegiance to the community, religion, taboos, emergence of primitive leadership, fear and panic, family roles, survival under conditions of stress, etc ...

Another way of considering disasters deals with their long-term effect on development. The Lisbon earthquake could be studied in relationship to the decline of the Portuguese Kingdom, and one is free to speculate on the association between living on the St Andreas fault and the blossoming of new religions.

Still another way is to handle disasters as a fac-

tor of mortality, morbidity, incapacity, loss of work,
that is a phenomenon interfering with the health of populations. This is epidemiology.

Now, one should ask, what is epidemiology, and what is
health? Epidemiology has been defined as "the study of
the dynamics of diseases in populations, in order to
identify their determinants and design ways for controlling them". In other words, and to speak in plain
terms, who gets hurt, how many, why, and what to do.

The definition of health proposed by WHO is, as
everybody knows, a challenging but somewhat inoperation-
al one: "Health is a state of complete physical, mental
and social well-being, and not merely the absence of
diseases or infirmity".
It follows that natural disasters in general, including
earthquakes, are working material for the epidemiologist, just as cholera, coronary, suicide or traffic accidents.
The basic questions that the epidemiologist is used to
asking himself with respect to disease or other health
nuisances are:
- how important is it (how many people affected)?
- what does it do to populations (death, incapacity,
  permanent disability, etc....)
- how does it operate (which are the factors involved?
  where, when and how often it occurs? in which population groups?);
- how to prevent it (identify, quantify and simulate
  the operating factors);
- how to control it.

Epidemiology is the basic tool of health planning.
The job of the epidemiologist is to measure phenomenons
and collect information in order to rank priorities,
design strategies and draw baselines for evaluation. If
reduction of deaths and incapacity from earthquakes
have to be achieved, these events have to be studied
along the same lines.

EARTHQUAKE AS A MORTALITY FACTOR. MAGNITUDE OF THE PROB-
LEM.

One might wonder whether earthquakes are an important factor of mortality. Where so? How do they compare with other mortality factors?

Accurate mortality figures for earthquakes are

generally hard to get. Census data are often lacking.
Counting of the dead may be impracticable or impossible.

For the period 1951-1968, the average number of
deaths from earthquakes was estimated at 3.650 per year.
This is probably less than deaths from snakebites, considerably less of course than deaths from such an easily preventible disease as tetanus, to say nothing of
deaths from car accidents or cigarette smoking, several
orders of magnitude greater.

On a historical basis, it was estimated that 10
to 15 million people lost their lives in earthquakes
during the historical period, i.e. the last 4,000 to
6,000 years. According to Latter,[1] approximately 5
million people were killed by earthquakes in the last
millenium. However large these figures might seem,
one may be tempted to say that earthquakes have killed
less people in the whole history of mankind than the
anopheline mosquito during the last pre-DDT decade.
They actually claim less victims each year than unprop-
erly marked pesticides mistaken for food by poor rural
populations.

What makes earthquakes important, however, is not
so much the size of the mortality in the long run, but
the fact that they happen all of sudden, are localized
in one place, and affect large numbers of people. In
other words, they are time- and place-clustered events.

For example, for the period 1951-68, the number of
deaths from earthquakes was estimated at 67,500. Yet,
about half this number occurred in Iran. The number of
annual deaths from earthquakes in the world for the
same period ranged from 110 (in 1961) to 21,000 (in
1960). The Managua, Nicaragua, earthquake alone killed
an estimated 2,000 to 5,000 on December 23rd 1972. It
has been calculated that, should an earthquake of the
same magnitude as the one which struck the Shanshi Province, China, in January, 1556, killing around 830,000,
occur again, it would cause over one and a half million
deaths.

Being sporadic and highly localized events, earthquakes, as most natural disasters, attract much attention from world opinion but very little from health experts and long-term planners. It is, however, by no
means clear whether death is a valid index to evaluate
the degree of preparedness for earthquake damages. The
mortality for a given earthquake depends on many variab-

les, among which are population density, adequacy of
structural design, stability of foundations, magnitude
of the earthquake, and time of occurrence. This last
factor is particularly important. It has been estimated that had the Alaska, 1964, earthquake occurred at
a different time, during the cannery season, and when
schools and public buildings were occupied, the number
of deaths would have been multiplied many times. This
makes population of the quake stricken area an invalid
denominator for calculating probability of death. In
countries with high seismic hazards, a nation-wide denominator will be more useful to calculate death-rates
over periods of several years. This could reveal interesting time trends with respect to the degree of
preparedness,or to the adequacy of building techniques
(especially inappropriate replacement of wood or adobe
by concrete) in rural areas.

The ratio of the number of injured to dead people
is another interesting index, being generally very
small. It ranges from 1 to 9 to 1 to 20 in Agadir
and in some recent earthquakes in Iran. The definition
of injuries may vary and has to be precisely defined.
However, such a ratio clearly points to the lethal effects of earthquakes on population. It stresses the
need for better prevention through adequate civil engineering measures. Such figures should be included in
cost-effectiveness analysis of prevention measures vs
relief and rehabilitation.

THE ECOLOGICAL DIMENSION OF EARTHQUAKES

To assess the importance of earthquakes by taking
large populations as a denominator of the rates, and/or
averaging over long periods of time, is an epidemiological fallacy. It might at times justify the low rank
priority given to natural disasters by planners and ad-
ministrators. It does not however preclude the mobili-
zation of world interest and the pouring of considerable resources of all kinds, often on a spontaneous and
non-coordinated basis. Three main epidemiological features of natural disasters in general, and especially
earthquakes, might explain why the world at large is so
much concerned: 1) they are sudden; 2) a large number
of people are affected at the same time; 3) they also
kill adults. This is a general observation in health
sociology. To move people, these three characteristics
have to be present. Home accidents kill many more
prople than plane crashes but never make headlines. In

a number of countries, infant mortality is responsible
for the loss of more years of life-expectancy than all
cancers together, but these deaths are in children who
are easy to replace. Cholera is a good example. Its
spread throughout Europe in the nineteenth century,
though much less devastating than alcoholism, tubercu-
losis, diptheria or a number of occupational diseases,
nevertheless triggered the whole development of public
health. It was occurring suddenly - gone to work in
the morning, back to the cemetery in the evening - ;
it affected large numbers of people, - London 1854,
Hamburg 1892 - ; and it killed adults, from Charles X
to Tchaikowsky, a quite embarrassing affair. It was
responsible for the first convention on health statistics, the creation of the Pan American Sanitary Bureau,
and many other developments in public health.

Sporadic and highly localized events such as major
epidemics or earthquakes assume an ecological dimension.
The normal living conditions are being put off balance,
the number of deaths and damage exceeding the capacity
of absorption of the community.

There are stresses and strains in the usual daily
life: diseases, death in the family, interpersonal
conflicts, social and material deprivation, failure of
expectations of various sorts. But these are individual crises and the sufferer has to find his own solution
for which the social group has generally developed
models of acceptable behaviour. There are socially recognized ways of handling an incurable disease, death
of a relative, business failure, or smashing the car.
These individual disasters do not induce change in the
society as long their incidence is random. Simultan-
eous occurrence, as in disasters, may destroy the ecological and social balance, the effects exceeding by far
what is expected from the number of individuals suffer-
ing. Such a situation, in which the needs exceed the
capacity of the community, calls for external emergency
assistance.

This, incidentally, answers the long-debated and
somewhat sterile question of defining a disaster. The
criterion is the need for emergency assistance, as
stressed in a NATO definition: "a disaster is an act of
nature or an act of man, which is or threatens to be of
sufficient severity and magnitude to warrant emergency
assistance".

DISASTER CULTURE

However, capacity to absorb a disaster may vary
considerably according to the community concerned.
Some populations with high exposure to disaster develop
some kind of social adaptation. This has been described
as disaster culture. Such are populations with repeated
experience of typhoons in Pacific islands (Yap).[2] The
same was true of mining towns in the nineteenth century
or, on a more localized scale, fishing communities.

This should be studied in the light of the perception of natural hazards by various populations: why are
people living on the slope of a volcano, why are they
rebuilding a city on the exact location of repeated
earthquakes, how aware are they of the risk, why do
they refuse to move? This has been the subject of considerable investigation in recent years. With respect
to earthquakes, there is a large theoretical range of
possibilities, such as moving dwellings to other sites,
soil and slope stabilization, sea wave barriers, fire
protection, anti-seismic building design, appropriate
warning systems, scheme for emergency evacuation, sub-
sidized insurance, etc... However, as Burton[3] points
out, ["men persist in hazard areas and continue to occupy them at an increasing rate in spite of the sure
knowledge that disasters will certainly follow. A not
entirely obvious explanation is that men often have
good reason to be there. Hazard areas may present
economic opportunities superior to those available else-
where, at least from an individual point of view, or to
a non-professional perceiver. In some cases, people
located in hazard areas would find it extremely difficult to move out without help from some outside sources,
or to do so may require sustaining a loss, the abandoning of an investment or a livelihood. Even when the
reasons for being in a hazard area are not absolutely
compelling (as in the use of seacoasts for vacation
homes and recreational amenities), many people do not
feel strongly threatened. In such cases they may eliminate the hazard from their perceptions, or reduce them
to some manageable and comfortable level. Even if
people have no good reason to be in hazard zones and do
feel threatened institutional arrangements in society
often operate to keep people in the same place and to
protect existing interests by reinforcing the status
quo".

The study of the psychology of disaster-exposed
communities is of major importance for the pre-disaster

planning and organization of the whole forecast and
warning system.

PREVENTION, PREDICTION AND WARNING

As often repeated, prevention is better than cure.
One generally adds that it is cheaper to prevent than
to cure, which is not always proved.

The whole problem of prevention - prediction -
warning, be it for natural disasters or diseases, should
be considered as what it is, i.e. a highly complex
system which involves a sequence of decisions based on
probabilities for which statistical observations are
scarce.

Social and health effects of natural disasters
will depend as much on proper prediction mechanics as
on the magnitude of the disaster itself. It is clear
that action at this level should be sought, for it will
considerably increase efficiency of pre-disaster planning.

Earthquakes are positively different from other
disasters in this respect, as their occurrence can only
be computed as a probability over a long-term period.
Maps can be drawn with the respective frequencies of
earthquakes, defining high-risk areas. Apart from this,
there is still little or no forecast, and therefore
generally no possibility for early warning.

For most other disasters, such as floods, or even,
on a different time-scale, typhoons and tsunamis, prediction can be made on the basis of pre-event signals,
making prevention possible either by removal or by
early curative measures, just as early diagnosis im-
proves the prognosis for a number of chronic diseases.
Earthquakes are more like accidents. You know of dangerous roads and special hazards, but when the crash is
eminent, it is too late, or at least preventive measures, such as crash-safe cars, are technologically sophisticated and expensive.

For any type of disaster, the relationship between
prediction and prevention can be represented in the
following graph (Figure 1).

For the sake of simplicity, one may consider five
successive steps.

292

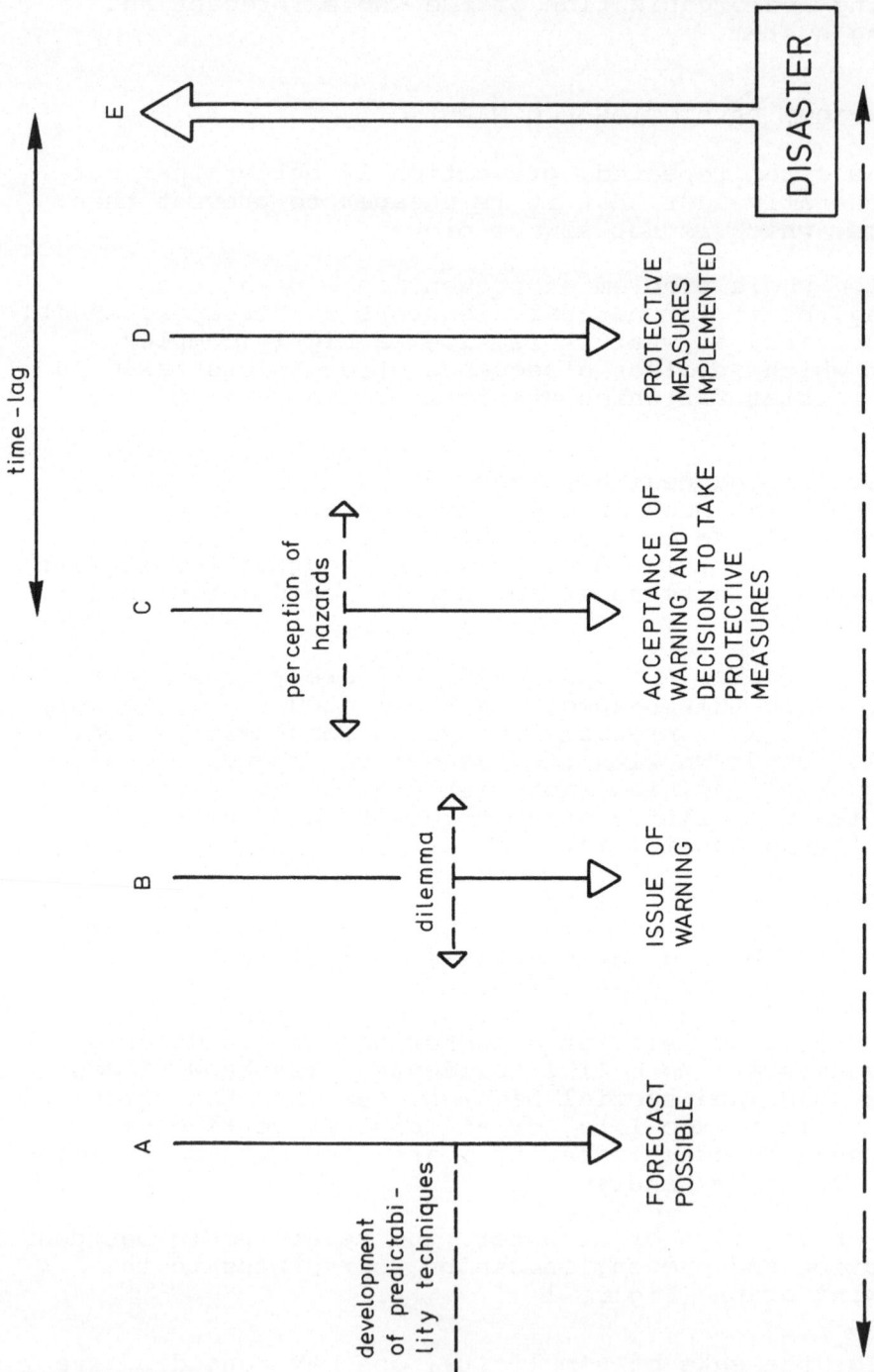

Figure 1: Relationship Between Prediction and Prevention of Disasters

a) <u>Forecast possible</u>. This is the time at which, according to present technology, forecast of a disaster is possible. The development of more sensitive forecast devices requires heavy expenditures in research.

b) <u>Issue of warning</u>. This is the time at which notification of a possible disaster is relayed to the population. This implies a decision balancing the respective risks of giving a false alert or increasing unpreparedness by postponing warning.

c) <u>Acceptance of warning</u>. This is the time at which the population accepts the warning and decides to move to take preventive measures. It will to a considerable extent vary according to the manner of warning given, previous disaster experience, perception of natural hazards by the population, and education. There has been a large amount of interesting study on this topic. It is important to view warning as a total process of communication and organization, and not only as the dropping of a message to which people do not react. It moreover supposes that the population exposed to disasters knows in advance which measures are effective.

d) <u>Having taken protective measures</u>. This depends on the time required to implement protective measures. It varies with the type of disaster, from running miles away in case of volcanic eruption to opening the back doors and staying away from the windows in case of tornadoes.

e) <u>Disaster</u>. The time lag between the decision for taking protective measures and the moment the disaster strikes will determine the efficiency of protective measures. If this is less than the time required for protective measures to be implemented, prevention will be reduced or nil. In many instances casualties will even be worse than expected if no warning was issued, because the process of taking measures (such as shutting the windows, climbing the stairs or running in the streets), is often likely to increase exposure to disaster-associated hazards.

For earthquakes, warning is coincidental with the first tremor. If this is slight, and prodromal to stronger quakes, some protective measures can be taken. There are however no rules. In the Managua earthquake, 1972, an earth tremor was felt throughout the city at 9.30 p.m., on December 22, more than three hours before the major quake (December 23, at 12.39 a.m.).Few people took any precautions.

PHYSICAL EFFECTS OF THE IMPACT AND COMMUNITY REACTIONS

Immediate Reactions

Immediate physical effects of the impact will depend on the magnitude of the earthquake, type of building, density and distribution of the population, time
of day, and possible prodromal signs. Mortality rates
may vary considerably for disasters of similar magnitude, from 7 per 100,000 (Banja-Luka, 1969) to over 3
percent (Managua, 1972). The ratio of amount of damage
to properties to number of deaths is generally high for
earthquakes, particularly in developed countries
(960,000 US $ of damage per life lost in the Fernando,
Calif., 1971, earthquake).

Impact effects include immediate ones, such as being killed or wounded by collapsing buildings, and delayed ones, such as being buried or trapped under debris. This calls for immediate rescue, the chances of
survival decreasing rapidly with time. Rescue cannot
generally wait for external assistance, especially in
remote areas. One important consequence is that appropriate self-mobilization of the survivors will be decisive in reducing the number of lives lost.

There has been a fair amount of information gath-
ered on the reaction of disaster-struck populations.
Sociologists have shown that many of the most immediate
and pressing tasks of rescue and relief are accomplished by the survivors themselves. Although most persons
subjected to sudden violent disasters are momentarily
stunned and confused, they usually, within a short time
regain enough self-control to extricate themselves, if
they are physically able, and to assist relatives,
neighbours and immediate associates (Fritz[4]). A number
of studies have shown that within half an hour, a large
proportion of the survivors, up to 75 percent in some
disaster, are engaged in some kind of rescue activity.
This, however, varies according to cultural patterns.
Within a given population, delay of recovery from the
post-impact atonement and proportion of people participating in community rescue, vary according to different
groups. The relative lack of disaster relevant skills
in highly technologically developed societies is noticeable as compared with disaster-culture communities.
Not only is a community relying heavily on a highly elaborate technology more exposed to large-scale break-
downs of vital utilities, but its members are also gen-

erally highly differentiated, making them incapable
of facing any disruption of the ecology. As it was
said "the more one goes to school, the less one knows
of current dayly life". Everything becomes the job of
an expert, from replacing a fuse, to cooking an egg, to
sewing, to taking care of minor daily frustrations even
to talking to each other, so what about crawling, lift-
ing, running or extricating people from debris! In
disaster-cultures, sucs as was described in Yap, a
typhoon-exposed atoll in the Pacific, everybody is a
fireman, a policeman, a doctor. In our cultures,
people are trained to have skills in one particular
highly specific area and to be unskilled in all the
rest.

This, incidentically, is important with respect to
using foreign volunteers for disaster relief. Within
a particular community, the proportion of people ready
for self-help varies according to various groups. It
has been shown that people with some disaster skills
are much faster to recover and participate in rescue
activities.

An important implication is that training for rescue and first aid in disaster-prone countries, such as
it is organized by national Red Cross/Crescent Societies, will be decisive in saving lives in the period
immediately following the impact. On the other hand,
various categories of professional personnel in these
countries, from agricultural instructors to administra-
tors, should receive a basic training of this kind, or-
iented towards the most common types of natural disasters likely to occur in their area.

However, self-help in many instances is reduced or
delayed by what sociologists have termed role conflict.
In emergency situations, different groups in which
people have roles may require participation at the same
time, creating a role conflict for the individual, and
a shortage of personnel for some groups. Role conflicts are illustrated by the fireman who first checks
on his family, the policemen who instead of reporting
for duty rushes to protect his relatives, the doctor
who first phones home before joining a rescue team.This
was first observed during the famous Halifax explosion,
in 1917, when it was reported that as a rule people
with families ran first to their homes in order to find
out if their own relatives were in danger. Many well
-documented case-histories report similar observations,
several relating particularly to earthquakes in Japan.

296

Although it might seem in contradiction with the pub-
licized ethics of various professional groups, this behaviour is only normal and should be taken into account.
For the sake of increasing the efficiency of rescue and
relief-activities, provision should be made for solving
role conflict by giving facilities to people to contact
their relatives as soon as possible. Several studies
have shown that people with duties towards the community can also perform these duties successfully after
they have had an opportunity to check on their own families - whatever the results of such a check might be.
Whether the relatives are safe, hospitalized or dead,
the important thing is that the family responsibility
has been assumed and information received. It should
be noted that role conflict has been strongly question-
ed by some sociologists. For White,[5] role conflict implies not so much family as familiarity. People choose
to perform along the lines with which they are most familiar. For most people this means taking care of relatives, or for clerics concentrating on some parish-
ioners, whereas for other professional groups it could
be carrying on well-defined usual tasks. She inter-
viewed a number of people involved in tornadoes in the
U.S.A., and members of disaster-relevant organizations,
such as city government officials, police, fire-depart-
ment, power and gas companies, radio stations, hospitals, Red Cross, Salvation Army and Civil Defense. This
lead her to suggests that under unique conditions of
stress, characterized by uncertainty as to the action
to be taken, the individual with role conflicts will
try to solve them by grasping at what she calls the
"first certain solution". By "first certain solution"
is meant the first course of action for which (1) the
problem is clearly defined; and (2) there is something
the individual can do to help solve the problem. In
other words, "the individual will jump at the first
chance to do something to help".

This is in line with what was said previously
about the increased efficiency of people having skills
of some kind. Training for rescue in disaster-prone
communities will at the same time increase motivations
of people to help by providing them with skills, and
contribute to reduce role conflict by making rescue
tasks more familiar. This is confirmed by observations
of utility workers involved in disasters, whose efficiency has been shown to be particular high.

A last point is the visibility of the position. In
small communities struck by a disaster, it was observed

that persons with unusual social positions, known to
everybody and expected to perform their professional
roles, were obliged through social pressure to exert
their responsibilities. This might explain the useful-
ness of uniform and visible insignia of one type or another.

Post-impact Medical Care

Once people have been extricated from the collapsed buildings, the next problem is taking care of the in-
juried. In contrast with other types of disasters,such
as floods, where medical care is often not a priority,
earthquakes represent the major indication of emergency
health techniques under natural conditions, i.e. ex-
cluding man-made disasters.

From scattered data and common experience, there is
clear evidence that the incidence of physical body damage is high, although the literature is scanty on the
relative frequency of the various lesions. Bone fract-
ures and laceration of joints are usually reported,with
crushing, cutting or tearing away of muscles and lesions
of the skin. Open wounds are common. It has been reported that among survivors 40 percent have minimal lesions or do not require medical attention, 40 percent
present injuries whose treatment can be delayed, and 20
percent require immediate life-saving care. This probably refers to earthquakes of high magnitude in densely
populated areas, for these figures seem particularly
high.

More data should be collected on the type of lesions observed, as associated with the types of building
materials in order to design and standardize emergency
rescue kits. This information could also be useful for
drawing building regulations in earthquake-prone areas,
with a view to reducing certain types of body damages.

Collecting such data in an emergency situation is
not easy, but this should be achieved if the number of
victims is to be reduced through efficient predisaster
planning. Rescue and medical care in earthquakes
should strictly adhere to the basic principle of emergency medicine for the management of mass-casualties:
moving, sorting, and using standardized procedures.This
differs radically from day-to-day medical practice.
"Under ordinary circumstances, the profession goes all
out for all patients at all times. In disasters, it is
foolishly cruel to do so because such economic use of

the resources denies life-saving measures to many victims."

Time, personnel and resources are grossly insuf-
ficient to meet all the needs. In such circumstances,
sorting is therefore an indisputable responsibility for
anyone who helps, to make his aid available to those
who can really benefit the most from it. Very often,
this means caring for the slightly injured, to the de-
triment of the badly injured for whom little can be
done. Defining priorities and making a choice according to the chances of success is something many medical
people are most reluctant to do. Current medical
training glorifies individual care, selecting the most
affected case. This, however, is something to avoid in
disaster situations. It is high time that medical
schools made place for a couple of hours training on
the basic principles of mass-emergency care. This is
especially important in view of the assault of volunteers, including doctors, in disaster areas. This assault, which results from the spread of information,
improved communications and the drive to accumulate all
kinds of personal experience, is likely to increase in
spite of all the recommendations from disaster relief
professionals.

Standard procedures are especially important in
view of the fact that channelling the patients through
various services, from outposts to hospital, and multi-
plicity of medical personnel, precludes to some extent
individualized care. Continuity of care under team
work conditions is possible only when procedures are
standardized.

The high incidence of injuries following earthquakes, together with the immediate physical effects,
such as burial or trapping, has another important consequence. Medical personnel, when on the spot, cannot
restrict themselves to highly skilled tasks. The aim
is to save as many people as possible, not merely to
assume professional roles. This might mean handling
picks or shoveling debris. Thus surgeons and other
medical personnel participating in rescue operations
should also, whenever necessary, help to extricate and
move victims.

What disasters most need is polyvalent people,that
is people with many non-specialized skills. What most
assistance consists of, on the contrary, is highly
specialized people (including doctors) plus often tot-

ally unskilled volunteers.

A logistics for the rescue and medical aid of populations in earthquake devastated areas has been designed in the Los Angeles, Calif., area, after the 1971 earthquake. It involves moving medical teams by helicopters from hospitals outside the damaged area to sort the patients and give first aid, while the nearest hospital are mobilized for receiving the victims.

Delayed Health Effects of the Impact

People successfully rescued, with no major lesions after having been buried under debris for several hours, are apt to develop crush syndrome within a few hours. This is a syndrome characterized by swelling, paralysis and insensitiveness of the compressed extremities.Weakness, sweating, thirst, pallor, low blood pressure are part of the syndrome. There is progressive restriction in the amount of urine passed (oliguria, anuria). The overload of the circulating blood with resorption products (myoglobine) from the crushed muscles plays a major role in the pathogeny of the crush syndrome. Prognosis is bad, death occurring in up to 65 percent of the people with crush syndrome according to some statistics. Renal dialysis decreases this fatality rate considerably. However, in most earthquake conditions, there are no renal dialysis facilities around, and in any way not to an extent sufficient to handle mass-casualties. Crush syndrome was reported in 16 percent of the injured after the Agadir earthquake. It raises tantalizing decision dilemmas. It occurs in people treated successfully for shock. The faster and the better blood circulation is re-established in a shocked survivor, the more chances he has to develop the syndrome. An efficient way to prevent it would be immediate removal of the affected limb(s).

Another delayed hazard from earthquakes is fire. Training for earthquake rescue and first aid should include the treatment of burns, which is a very delicate problem.

Accidents to rescue volunteers and other helpers should not be overlooked. In the Anchorage, Alaska, 1964 earthquake, it was reported that 20 percent of hospital-treated injuries were accidents occurring after the quake in persons working with debris.

PSYCHOLOGICAL AND SOCIAL EFFECTS OF DISASTERS

The psychological and social effects of natural disasters on the individual survivors, injured or not, and on the community, are fairly well known. At any rate, more information seems to have been collected in this respect than in the health field.

Impact and Post-impact Reactions in the Individuals

From the evidence available, the impact reaction in earthquakes and in some other natural disasters, such as volcanic eruptions, is flight. The patterns of flight are family-clustered. Retrospective studies have showed that generally the family members present when the danger was perceived flee together, people searching or coming back to make sure that nobody is left behind. This may last until the impact is through or people are trapped by collapsed structures.

The post-impact response in individuals is described as "disaster syndrome" (Wallace[6]). It consists mainly of three phases:

1. First, there is what has been reported as stupor. People sit motionless, often with little awareness of the extent of damage to self, family and community. Efforts at first aid, rescue and evacuation are inadequate or absent. This state is considered as a pseudo-infantile level of adaptive behaviour. It may persist for minutes or hours. As might be expected, it is more frequent, and has a longer duration, in injured people as compared to non-injured survivors.

2. Later, the pattern changes. Uninjured survivors are docile, ready to help, grateful, and extremely anxious to hear that others have survived too. This stage of recovery corresponds to an identification of the individual with the community. As mentioned, a larger number of survivors are engaged in rescue and relief activities. This sometimes coincides with the arrival of external rescue workers.

3. It is followed by a stage of euphoria and intense public spirit. Structures symbolic of community solidarity, such as churches, are cleaned up and repaired. There is eagerness to work for community welfare. Family ties are strengthened. Social and ethnic barriers are ignored. Neighbours help each other and people share resources. At the same time, considerable

complaint is directed towards external relief organizations; people resenting dependence as interfering
with their drive to reassume the responsibility in
their community. As pointed out by Quarantelli [7]:"to
the locals, it is "their disaster" and they do not want
any outsiders coming in to take credit for "their" work
during the emergency period".

The two main points to stress are

1/ the considerable amount of resources for self
-help found in disaster-struck communities. One should
therefore insist on better preparedness through adequate training, and include community resources in planning for disasters;

2/ the necessity to give appropriate attention to
the reaction of the community when organizing external
assistance, in order to prevent conflicts between the
rescued population and the workers from outside. A
good knowledge of post-disaster psychology will facil-
itate and increase the efficiency of relief. Stress in
the survivors is likely to be considerably increased by
evacuation and temporary or permanent relocation in un-
natural or at times hostile environments.

In the long run, psychological and emotional problems may become a first priority for assistance. This
might be especially important when external assistance
is removed and during the early phase of the so-called
rehabilitation (or further deterioration?) period.

Social Response of Disaster-struck Community

The social response will vary with the experience
the community has of disasters.

Three points should be mentioned: the general absence of panic convergence behaviour, and the emergence
of new leadership. These are important for adequate
planning.

Sociologists have shown that more often than not
panic is a myth. Panic, defined as a contagious process of uncontrolled mass-fear, seems restricted to situations in which people are trapped in a closed space
with no decision to take and no alternative behaviour
available. Such would be the situation in a bunker, or
in a building in fire with no exits.

Convergence behaviour refers to all movement towards the disaster area. It includes physical movement of persons motivated by curiosity, interest or drive to help, incoming messages of information and offers of help, and unsolicited equipment and supplies. It continues for days or weeks, extending in successive waves from the periphery of the disaster area to far-away places or foreign countries. It may block the means of transport, overload local warehouses, heavily tax the scarce resources left with thousands of volunteers, and jam communication and information facilities.

At times, this might result in what could be called a second disaster, man-made, added to the initial one. Effective control of the convergence process requires provisions for adequate management at the pre-disaster planning stage. As pointed out by Fritz,[4] it should not be narrowly restricted to blocking or restraining it. Convergence is often "motivated by anxiety over missing kin and friends, symphathy for the stricken population, and the desire to help it. For these needs to be satisfied, disaster management must provide adequate information, positive direction, and guidance, rather than indiscriminate restraint".

The way new leaders emerge in disaster-struck communities should be studied further. It is briefly mentioned here in order to stress the importance in organizing relief of identifying true leaders and efficient channels of authority in the community.

A striking observation is the way human communities at large are apt to cope with and recover from disaster. Fritz points out how, to an external observer entering a disaster area, the behaviour of survivors might seem irrational. For this observer, survivors will seem subject to completely irrational, chaotic, aimless, random, uncontrolled or conflicting activities. "People are running or driving vehicles in opposite directions, often passing each other without acknowledgment or seeming awareness. Some persons are moving out of the impact area, many others are moving in. Others are standing around just looking or talking to each other. Here and there small groups of people are digging in debris, comforting the injured, or attempting to retrieve their scattered belongings. Behaviour is so mixed and conflicting that it appears to have no rational pattern .... (This is) not panic, but social disorganization - that is, uncoordinated activity on a general, community-wide or societal level. It is important to recognize that this social disorganization need not indicate personal or small-group disorganization. On the contrary, many individuals and small groups are acting with purpose and considerable control to cope with problems they face; however, their attention is apt to be focused on the immediate tasks at hand and so they appear oblivious to the more general needs that have arisen as a result of disaster."(Fritz[4])

This pattern of seemingly aimless activities is characterized by low efficiency but probably at the same time high effectiveness. People in disasters discover new ways of performing things. This could represent some type of adaptive behaviour to cope with disasters, social disruption and primitive ecological conditions. In some obscure way, post-disaster behaviour in man seems to have a selective value for survival, not unlike the way the reaction of termites when their hill is damaged.

HEALTH PROBLEMS AFTER DISASTER

Following a disaster, populations are exposed to many health hazards. Urgent requests for health supplies and medical personnel are generally made. Health is high on the list of priorities, and may be affected by a variety of disaster-associated factors.

1. Loss of housing exposes the disaster-struck population to physical hardship such as cold. This may result in various ailments, such as pneumonia, especially in infants and aged persons. Tents, or other types of temporary shelters such as polyfoam igloos, are one of the most urgent supplies needed. Care should be taken in selecting the place where tentopolis or other survivors villages are to be located. Proper water supply and hygienic facilities must be provided if epidemics are to be prevented. Temporary shelters must withstand a change of seasons (for example a coming rainy season), and should not become permanent.

2. Except in special conditions (mountain sites isolated by landslides, islands or other places of difficult access, or quakes occurring in areas already affected by drought and famine), food shortage should generally be avoided, relief organizations arriving rapidly on the spot. Local conditions such as lack of transportation, administrative restrictions or collapse of the distribution system could, however, more than

the disaster itself, provoke food shortage. Shortage
of milk for feeding babies is of particular importance.
Caloric requirements of refugees, as well as the amount
of the various nutrients to be supplied, have now been
fairy well studied. Special attention should be given
to the health hazards associated with contamination of
food (for example dilution of skimmed milk with un-
boiled cholera-infected water).

3. Water-supply is commonly cut off or contamin-
ated. Sewers are leaking, possibly resulting in massive contamination by pathogenic gastro-enteric germs.
Germs of typhoid fever and bacillary dysenteric may
survive for weeks in sewage, salmonella even for months.
Overcrowding and poor sanitation in refugee camps
enhances the hazard of faeco-oral diseases, such as
typhoid, paratyphoid, bacillary dysentery and viral he-
patitis. If cholera-carriers are present in the community, an outbreak could occur. Control and repair of
waterworks is a high priority, as is monitoring of the
water-supply. Prevention of waterborne diseases in the
population requires a supply of clean water for drink-
ing and adequate sanitary facilities. Pumps, emergency
chlorination plants and disinfectants are among the
most pressing needs. The World Health Organization has
recently (1971) published a "Guide on Sanitation in
Natural Disasters" which should be consulted.[8]

4. As mentioned, overcrowding increases the incidence of diseases transmitted through faeco-oral contact. It also increases the incidence of direct contact diseases, such as smallpox and measles. Smallpox
should be considered as a potential danger in populations not adequately immunized. As long as the disease is not world eradicated, vaccination campaigns
should be organized when appropriate.

Measles is likely to be an important killing
factor in young children mixing in overcrowded surround-
ings.

5. Poisoning by industrial residues, including
radioactive residues, is a potential hazard. In cases
of associated floods, dairy product wastes or industrial sewage can be sources of contamination. Leaking
pipes or oil plants may result in an overflow of petrol-
eum products onto the surface of water.

6. Rodents are a reservoir of certain transmissible
diseases (plague, tularemia). Under given epidemiolog-

ical conditions, outbreaks of such zoonoses can occur.
Rats are particularly important reservoirs and should
eventually be destroyed. Rabies, mainly transmitted
by stray dogs, will also require appropriate measures,
according to local conditions. In the long run, earthquakes may increase transmission of vector-born diseases, by destroying or washing out the buildings
sprayed with residual insectisides. They may also multiply breeding places of insect vectors in the debris.

7. Carcasses of dead animals can occasionally be
a source of infection for communicable diseases and
should be destroyed. This hazard is less in earthquakes than in other natural disasters such as floods.
It should be stressed that putrifying bodies are not an
important source of disease.

8. Accidents to refugees living in non-familiar
conditions, or to persons scavenging the debris to rec-
uperate lost property, are likely to make additional
demands on the medical facilities.

9. Psychological stress and mental unbalance can
in the long run and as time passes become an important
factor of morbidity in the survivors. This is often
overlooked.

10. Special attention will be given to maintain-
ing medical care, especially the care of women giving
birth, including the prevention of neo-natal and post
-partum tetanus.

The general attitude to disaster is that it is
necessarily linked with epidemics. There is pressure
to vaccinate against all kinds of diseases, such as po-
liomyelitis, tetanus, typhus, typhoid, yellow fever or
cholera. Generally the risk is grossly exaggerated.
Epidemics are not inevitable following a disaster; and
mass immunization can severely hamper other efforts to
restore health care by diverting personnel and equipment. What is needed is a well-organized epidemiological surveillance system and this is a major requirement. It may prevent vaccinating large numbers of
people, giving a false safety, and help to stop outbreaks of communicable diseases among survivors and refugees. This monitoring system should be extended to
the use of medical facilities (number and location of
physicians, and nurses, location of supplies, etc...).

This review refers only to the most pressing
health needs in the disaster-struck population, provided the non-affected surrounding environment is normal.
If survivors, sometimes by the thousands or tens of
thousands, are taking refuge in areas with a precarious
economy and on the verge of collapse, the situation
will become still worse. The same remark applies when
the disaster-struck population is mass-evacuated to areas affected by severe drought and/or shortage of food.

This may result in man-made disasters, or social
conflicts, adding to the initial disaster.

Five remarks should be made with respect to health
relief:
1. public health problems should be replaced in
the whole context of disaster management, together with
communications, social needs, agriculture, public utilities, etc... There is no health assistance without a
well articulated logistics dealing with the various
problems of a disaster-struck community;

2. health is a word, and does not exist per se.
What does exist are specific health needs, which call
for specific items. There is too much of a tendency to
encompass all the health needs under simple headings
such as "doctors", "vaccines", "drugs", whereas doctors
might not be what the population most need in term of
health workers, there are vaccines for many different
diseases, and a number of drugs will be useless;

3. the health field should be demystified. Blood
and vitamins have magic connotations. They are not always what the population most need. Neuro-surgeons
might be less useful than truck drivers. Well equipped
surgical units may at times save less lives than field
kitchens;

4. useless medical supplies, including personnel,
are likely to jam transport and storage facilities like
any other useless item. That they are medical is no
excuse. Useless refers here not only to items which
are needless, but also to items which cannot be used
because they lack proper packaging or marking, it refers equally to personnel who do not have appropriate
training;

5. Medical and rescue workers should arrive well
-equipped and self-contained, with necessary supplies
for self-sufficiency.

It should be recognized that little has been done to study the epidemiology of disaster-struck communities according to various types of natural disasters and to systematize their health requirements. There is an urgent need for more research in the field of disaster epidemiology, in order to design efficient strategies for protecting health.

Finally, it should be emphasized that the rehabilitation period following earthquakes, when restoration of pre-disaster conditions is worked at, should be used to review health objective and distribution of health services. This is a good opportunity, if nothing else, to break what has been called the "cake of customs".

1. Latter, J.H., Natural Disasters, Advancement of Sciences, 362, 1969.

2. Schneider, D.M., Typhoons on Yap, Human Organization, 16, 10, 1957

3. Burton, I., Kates, R.W. & White, G.F., The human ecology of extreme geophysical events. Working paper No.1., Natural Hazard Research. University of Toronto, Department of Geography, 1968.

4. Fritz, C.E., Contemporary Social Problems, in Disaster, Harcourt, Brace and World, Inc., 1961.

5. White, M., In Community in Disaster, Barton, A.H., Double Day Ed., 1962.

6. Wallace, A.F.C., Mazeway Disintegration: The Individual's Perception of Socio-Cultural Disorganization, Human Organization, 16, 23, 1957.

7. Quarantelli, E.L. and Dynes, R.R., Image of Disaster Behaviour: Myths and Consequences, Disaster Research Center, Ohio State University, Working Paper N°37, mimeographer, 1971.

8. Assar, M., Guide to Sanitation in natural disasters, World Health Organization, Geneve, 1971.

LIST OF PARTICIPANTS

NATO ADVANCED STUDY INSTITUTE ON MODERN DEVELOPMENTS IN
ENGINEERING SEISMOLOGY AND EARTHQUAKE ENGINEERING:
IZMIR, TURKEY   JULY 2-13, 1973.

Scientific Directors:

J. Sólnes	professor	Faculty of Engineering and Science, University of Iceland Reykjavik.
E. Karaesmen	lecturer	Black Sea Technical University, Trabzon, Turkey.

Organizing Committee:

the two directors and

A. Aytun	gen. secretary	Turkish National Committee for Earthquake Engineering, Ankara, Turkey.
A. Gürpinar	civil eng.	Civ. Eng. Dept., Middle East Technical University, Ankara, Turkey.
S. Kavalali	civil eng.	Civ. Eng. Dept., Aegean University, Izmir, Turkey.

Lecturers:

N.N. Ambraseys	professor	Engineering Seismology Section, Imperial College of Science and Technology, London, U.K.

A. Aytun
N. Canitez   assoc.prof.   Institute of Earth-Physics, Istanbul Technical University, Istanbul, Turkey.

M. Caputo   professor   Instituto Di Geofisica, Universita Degli Studi, Bologna, Italy.

S. Cherry   professor   Civ. Eng. Dept., University of British Columbia, Vancouver, Canada.

A. Gürpinar
M. Ipek   civil eng.   Computing Science Dept., Istanbul Techn. University, Istanbul, Turkey.

M. Izumi   professor   Faculty of Engineering, Tohoku University, Sendai, Japan.

P.C. Jennings   professor   California Institute of Technology, Pasadena, California, U.S.A.

E. Karaesmen
M.F. Lechat   professor   Faculté de Médecine, Université Catholique de Louvain, Bruxelles, Belgium.

A.A. Moinfar   civil eng.   Technical Research and Standard Bureau, Plan and Budget Organization, Teheran, Iran.

K. Muto   prof. emeritus   Muto Institute of Structural Mechanics, Kasumigaseki Bldg., Chiyoda-Ku, Tokyo, Japan.

B.C. Papazachos   seismologist   National Observatory of Athens, Seismological Institute of Athens, Athens, Greece.

J. Penzien   professor   Earthquake Eng. Research Center, Univ. of California, Berkeley, Cal., USA.

H.C. Shah	assoc. prof.	Dept. of Civil Engineering, Stanford University, Stanford, California, USA.
J. Sólnes		
K.V. Steinbrugge	professor	Dept. of Architecture, University of California, Berkeley, California, USA.
S. Tezcan	professor	Civ. Eng. Dept., Bosporus University, Istanbul, Turkey.
Ö. Yilmaz	geophysicist	Turkish Petroleum Exploration Corporation, Ankara, Turkey.

Observers:

M. Dikmen	professor	Middle East Technical University, Ankara, Turkey.
K. Ergin	professor	Turkish National Scientific and Technical Research Found., Institute of Earth Physics, Istanbul Techn. Univ., Istanbul, Turkey.
Ö. Sayar	civil eng.	Izmir Technical Bureau, Ministry of Public Works, Izmir, Turkey.

312

Iran
E. Mardiross

civil eng., Eng. Seismol. Dept., Imperial College, London, U.K.

Japan
M. Tsugawa

civil eng., Muto Institute of Structural Mechanics, Tokyo.

Mexico
G. Lopez-Valades

civil eng., Institute of Engineering, National University of Mexico, Mexico.

Norway
E. Hjorth-Hansen

civil eng., Div. of Structural Mechanics, Norwegian Institute of Technology, University of Trondheim, Trondheim.

Pakistan
T.S. Wasti

civil eng., Civ. Eng. Dept., Middle East Technical University, Ankara, Turkey.

Portugal
V. Caiado (Mrs.)

geophysicist, Instituto Geofisico Do Infante D. Luis, Lisboa.

Sri Lanka (Ceylon)
A.P.S. Selvadurai

civil eng., Dept. of Civ. Eng., The University of Aston in Birmingham, U.K.

U.S.A.
A.D. Foutch

civil eng., Civ. Eng. Dept., University of Hawaii, Hawaii.

W.L. Heller

civil eng., U.S. Army Corps of Eng., Waterways Experiment Sta., Vicksburg, Mississippi.

J.O. Jirsa

assist. prof., Civ. Eng. Dept., University of Texas, Austin, Texas.

R.V. Sharp

geologist, U.S. Geological Survey, Menlo Park, California.

Turkey
H. Aktan

civil eng., Civ. Eng. Dept., University of Michigan, Ann Arbor, Michigan,USA.

G. Altinbas (Mrs.)

geophysicist, Institute of Earth Physics, Istanbul Techn. University, Istanbul.

N. Aydinoglu

civil eng., Civ. Eng. Dept., Istanbul Techn. University,, Istanbul.

V. Aykurt	professor, State Academy of Eng. and Architecture, Istanbul.
N. Bayülke	civil eng., Earthquake Research Dept., Ministry of Reconstruct. and Resettlement, Ankara.
I. Berktay	civil eng., State Academy of Eng. and Architecture, Istanbul.
G. Boser (Miss)	civil eng., Ministry of Public Works, Ankara.
S. Büyükasikoglu (Mrs.)	electr. eng., Institute of Earth Physics, Istanbul Technical University, Istanbul.
R. Can (Miss)	geophysicist, Imperial College of Science and Technology, Eng. Seism. Dept., London, UK.
M. Celebi	civil eng., Civ. Eng. Dept., Middle East Techn. University, Ankara.
H. Demir	professor, Civ. Eng. Dept., Istanbul Techn. University, Istanbul.
N.P. Erdem	professor, State Academy of Eng. and Architecture, Istanbul.
E. Ermutlu	civil eng., State Hydraulic Office, Research Dept., Ankara.
S. Gencoglu	geophysicist, Earthquake Research Dept., Ministry of Reconstruction and Resettlement, Ankara.
I. Gögüs	civil eng., State Academy of Eng. and Architecture, Istanbul.
U. Güslü	geologist, Institute of Earth Physics, Istanbul Techn. University, Istanbul.
A. Gul	geophysicist, Earthquake Research Dept., Ministry of Reconstruction and Resettlement, Ankara.
P. Gülkan	civil eng., Army Eng. Dept., Ankara.
F. Karadogan	civil eng., Black Sea Techn. University, Trabzon.
H. Karatas	civil eng., Civ. Eng. Dept., Istanbul Techn. University, Istanbul.

I. Kaya	civil eng., Civ. Eng. Dept., Aegean University, Izmir.
F. Keskinel	professor, Civ. Eng. Dept., Istanbul Techn. University, Istanbul.
D. Kolcak	geophysicist, Faculty of Science, Istanbul Techn. University, Istanbul.
Y. Konyalioglu	civil eng., Ministry of Public Works, Ankara.
K. Korkut	civil eng., Ministry of Public Works, Ankara.
M.A. Kuzu	civil eng., Ministry of Tourism, Ankara.
Ö. Küstü	civil eng., Civ. Eng. Dept., University of California, Berkeley, California, USA.
S. Özbakkaloglu	civil eng., Izmir Techn. Bureau, Ministry of Public Works, Izmir.
D. Özgür	civil eng., Civ. Eng. Dept., Middle East Techn. University, Ankara.
A. Özkan	civil eng., Civ. Eng. Dept., Bosporus University, Istanbul.
I. Sager	civil eng., Civ. Eng. Dept., Bosporus University, Istanbul.
C. Soydemir	civil eng., Civ. Eng. Dept., Middle East Techn. University, Ankara.
H. Soysal	geophysicist, Faculty of Science, Istanbul Techn. University, Istanbul.
A. Tabban	geologist, Earthquake Research Dept., Ministry of Reconstruction and Resettlement, Ankara.
N. Türkelli	geophysicist, Faculty of Science, Istanbul Techn. University, Istanbul.
E. Ülküdas	civil eng., Technical Services, Social Security Dept., Izmir.
Y. Wasti (Mrs.)	civil eng., Civ. Eng. Dept., Middle East Techn. University, Ankara.
G. Yavuz	architect, State Academy of Eng. and Architecture, Istanbul.